Ben Stacy Jerrik (Ed.)

Yelovsky District

AF708134

Ben Stacy Jerrik (Ed.)

Yelovsky District

Perm Krai, Russia, Krai, Subdivisions of Russia, Administrative centre

Part Press

Imprint

Publisher:
Part Press is a trademark of
International Book Market Service Ltd., 17 Rue Meldrum, Beau Bassin, 1713-01 Mauritius
Email: info@bookmarketservice.com
Website: www.bookmarketservice.com

Published in 2013

Printed in: U.S.A., U.K., Germany. This book was not produced in Mauritius.

ISBN: 978-620-1-56459-6

Contents

Articles

References

Yelovsky_District

Yelovsky District (Russian: Ело́вский райо́н) is an administrative district (raion) of Perm Krai, Russia; one of the thirty-three in the krai.[1] As a municipal division, it is incorporated as **Yelovsky Municipal District**.[2] It is located in the southwest of the krai. The area of the district is 1449 square kilometers (sq mi).[3] Its administrative center is the rural locality (a *selo*) of Yelovo.[1] Population: 10,743 (2010 Census);[] 13,121 (2002 Census);[] 14,127 (1989 Census).[] The population of Yelovo accounts for 49.7% of the district's total population.[]

History

The district was established on January 15, 1924.[3]

Demographics

The most numerous ethnic groups, according to the 2002 Census, include Russians at 94.1% and the Chuvash people at 2.6%.[3]

References

Notes

[1] Law #416-67

[2] Law #1761-364

[3] Encyclopedia of Perm Krai. Entry on Yelovsky District (http://enc.permculture.ru/showObject.do?object=1803751853&idParentObject=1803664521) (Russian)

Sources

- Законодательное собрание Пермской области. Закон №416-67 от 28 февраля 1996 г. «Об административно-территориальном устройстве Пермского края», в ред. Закона №84-ПК от 28 августа 2012 г. «О внесении изменений в отдельные Законы Пермской области, Коми-Пермяцкого автономного округа». Вступил в силу с момента опубликования. Опубликован: "Звезда", №38, 12 марта 1996 г. (Legislative Assembly of Perm Oblast. Law #416-67 of February 28, 1996 *On the Administrative-Territorial Structure of Perm Krai*, as amended by the Law #84-PK of August 28, 2012 *On Amending Various Laws of Perm Oblast, Komi-Permyak Autonomous Okrug*. Effective as of the moment of publication.).
- Законодательное собрание Пермской области. Закон №1761-364 от 10 ноября 2004 г. «Об утверждении границ и о наделении статусом муниципальных образований Еловского района Пермского края», в ред. Закона №175-ПК от 7 марта 2013 г. «О внесении изменений в отдельные Законы Пермской области, Коми-Пермяцкого автономного округа об утверждении границ и о наделении статусом муниципальных образований Пермского края». Вступил в силу через десять дней со дня официального опубликования. Опубликован: "Бюллетень Законодательного Собрания и администрации Пермской области", №12, I часть, 9 декабря 2004 г. (Legislative Assembly of Perm Oblast. Law #1761-364 of November 10, 2004 *On Establishing the Borders and on Granting the Status to the Municipal Formations of Yelovsky District of Perm Krai*, as amended by the Law #175-PK of March 7, 2013 *On Amending Various Laws of Perm Oblast, Komi-Permyak Autonomous Okrug on Establishing the Borders and on Granting the Status to the Municipal Formations of Perm Krai*. Effective as of the day which is ten days following the day of the official publication.).

Perm_Krai

Perm Krai	
Пермский край (Russian)	
— Krai —	
 **Flag**  **Coat of arms**	
Anthem: *none*	

Coordinates:	
Political status	
Country	Russia
Federal district	Volga[1]
Economic region	Urals[2]
Established	December 1, 2005[3]
Administrative center	Perm
Government (as of August 2010)	
- Governor[4]	Viktor Basargin[5]
- Legislature	Legislative Assembly[6]
Statistics	
Area (as of the 2002 Census)[]	
- Total	160600 km^2 (sq mi)
Area rank	24th
Population (2010 Census)[]	

- Total	2635276
- Rank	17th
- Density[7]	16.41 /km^2 (/sq mi)
- Urban	75.0%
- Rural	25.0%
Time zone(s)	YEKT (UTC+06:00)[8]
ISO 3166-2	RU-PER
License plates	59, 81, 159
Official languages	Russian[9]
Official website [10]	

Perm Krai (Russian: Пе́рмский край, tr. *Permsky kray*, IPA: [ˈpʲermskʲɪj kraj]) is a federal subject of Russia (a krai) that came into existence on December 1, 2005 as a result of the 2004 referendum on the merger of Perm Oblast and Komi-Permyak Autonomous Okrug. The city of Perm is the administrative center. Population: 2,635,276 (2010 Census).[]

Komi-Permyak Okrug retained its autonomous status within Perm Krai during the transitional period of 2006–2008. It also retained a budget separate from that of the krai, keeping all federal transfers. Starting in 2009, Komi-Permyak Okrug's budget became subject to the budgeting law of Perm Krai. The transitional period was implemented in part because Komi-Permyak Okrug relies heavily on federal subsidies, and an abrupt cut would have been detrimental to its economy.

Geography

Perm Krai is located in the east of the East European Plain and the western slope of the Middle Ural Mountains. 99.8% of its area are located in Europe, 0.2% - in Asia.

- length from north to south - 645 km;
- length from west to east - 417.5 km.

The krai borders with the Komi Republic in the north, Kirov Oblast in the northwest, the Udmurt Republic in the southwest, the Republic of Bashkortostan in the south, and Sverdlovsk Oblast in the east.

The krai borders stretch for over 2200 kilometers (mi). The highest point is mount Tulymsky Kamen (1,496 m).

Rivers

Rivers of Perm Krai belong to the Kama River Basin, the largest tributary of Volga River. There are more than 29 thousand rivers in Perm Krai. The total length of all rivers is more than 90000 kilometers (mi).

Chusovaya River

Only two rivers in Perm Krai have lengths exceeding 500 km. They are Kama River (1805 km (1120 mi)) and Chusovaya River (592 km (367 mi)).

There are about forty rivers with length from 100 km (62 mi) to 500 km (310 mi). The longest of them are:

- Sylva River — 493 km (306 mi)
- Kolva River — 460 km (285 mi)
- Vishera River — 415 km (258 mi)
- Yayva River — 403 km (250 mi)
- Kosva River — 283 km (176 mi)
- Kosa River — 267 km (165 mi)
- Veslyana River — 266 km (165 mi)
- Inva River — 257 km (159 mi)
- Obva River — 247 km (153 mi)

There are also many small rivers, but some of them have historical significance, for example Yegoshikha River, in mouth of which city Perm was founded.

Vishera River

Climate

It has continental climate. Winters are long and snowing. Average temperatures in January vary from −18 °C (− °F) in the northeast part of krai to −15 °C (°F) in southwest part. Record low is −53 °C (− °F) (in the north).

Minerals

Perm Krai is rich with minerals, that can be explained its diverse relief in mountainous and flat parts. There are produced: oil, natural gas, gold, diamonds, chromites, peat, limestone, building materials and others.[11]

Salt dump in Solikamsk

Oil in its area was first discovered in 1929 near settlement Verhnechusovskie Gorodki. Currently there are known more than 180 oil and gas fields. Among them are developed: 89 oil, 2 gas and 18 both oil and gas fields. Most of them are small and extracted in central and southern districts of krai. Northern fields a less developed because of deep lying of oil under salt layers.

Coal has been mined in Perm Krai for more than 200 years. For a long time it played an important role in the fuel and energy balance in the region. Maximum mining was in 1960 and reached 12 million tones, after it mining decrease and there are no exploration of new fields.

In Perm Krai is located Verkhnekamskoye deposit of potassium salts, one of the largest in the world. Its area is approx. 1,800 km² and the thickness of the salt layers reaches 514 m.

Kungur Ice Cave

Flora and fauna

Forests cover about 71% of krai's area. Predominant are coniferous forests, percentage of deciduous forests increase from north to south. There are 62 species of mammals, more than 270 species of birds, 39 species of fishes, 6 species of reptile and 9 species of amphibians.

Located in Perm Krai are 2 nature reserves: Basegi and Vishera.

Politics

Perm Krai Legislative Assembly, July 2010

During the Soviet period, the high authority in the oblast was shared between three persons: The first secretary of the Perm CPSU Committee (who in reality had the biggest authority), the chairman of the oblast Soviet (legislative power), and the Chairman of the oblast Executive Committee (executive power). Since 1991, CPSU lost all the power, and the head of the Oblast administration, and eventually the governor was appointed/elected alongside elected regional parliament.

The Charter of Perm Krai is the fundamental law of the region. The Legislative Assembly of Perm Krai is the province's standing legislative (representative) body. The Legislative Assembly exercises its authority by passing laws, resolutions, and other legal acts and by supervising the implementation and observance of the laws and other legal acts passed by it. The highest executive body is the Oblast Government, which includes territorial executive bodies such as district administrations, committees, and commissions that facilitate development and run the day to day matters of the province. The Oblast administration supports the activities of the Governor who is the highest official and acts as guarantor of the observance of the oblast Charter in accordance with the Constitution of Russia.

Administrative divisions

Welcome sign at the border of Chernushinsky District

Administratively, the krai is divided into thirty-three districts, fourteen cities of krai significance, and one closed administrative-territorial formation. Six administrative districts are grouped into Komi-Permyak Okrug, which is an administrative unit with special status formed within Perm Krai as a result of the 2005 merger of Perm Oblast and Komi-Permyak Autonomous Okrug, both of which used to be the federal subjects.

Municipally, the territories of all administrative districts and those of nine cities of krai significance are incorporated as municipal districts. The remaining five cities are incorporated as urban okrugs.

Demographics

According to the 2010 Census, the population of Perm Krai is 2,635,276;[] down from 2,819,421 recorded in the 2002 Census,[] and further down from 3,099,994 recorded in the 1989 Census.[]

Kamskaya Hydroelectric power plant, Perm

Ethnic groups, as of the 2010 Census, are:[] Russians (87.1%), Komi-Permyaks (3.2%), Tatars (4.6%), Bashkirs (1,3%), Ukrainians (0,6%), Udmurts (0,8%), Belarusians (0,3%), Germans (0,3%) and others. Additionally, 119,538 people were registered from administrative databases, and could not declare an ethnicity. It is estimated that the proportion of ethnicities in this group is the same as that of the declared group.[12]

There are about 40,740 Bashkirs is Perm Krai, according to the 2002 Census. Most of them live in Bardymsky District, in basin of Tulva River and belong to tribe gaina.

Tatars live in almost all settlements of Perm Krai. There are several different ethnographical groups of Tatar people. In this territory for long time was active contacts between Tatars and Bashkins, so in some cases it's difficult to delineate this ethnic groups, especially in such areas as Kuyedinsky District and Tulva River basin.[13] [14]

Vital Statistics for 2007: Source

- Birth Rate: 12.05 per 1000
- Death Rate: 15.70 per 1000
- Net Immigration: -1.0 per 1000
- NGR: -0.37% per Year
- PGR: -0.48% per Year

Vital statistics for 2012

- Births: 38 847 (14.8 per 1000)
- Deaths: 37 278 (14.2 per 1000) [15]
- Total fertility rate: 1.90(e)

Note: Data for Total fertility rate (2012) is estimate based on age and sex structure of Perm Krai at the beginning of 2012, number of births in 2012 and fertility structure in previous years.[16] [17]

Death rates in some of the remote and rural areas in Perm Krai are very high, never seen before during times other than major wars or natural calamities. Just five districts out of a total of 47 have a surplus of births over death in Perm Krai. The birth rate in Perm Krai is much higher compared to other European regions. For example, the birth rate for Germany was 8.3 per 1000 in 2007. Perm as a whole is having 50% higher birth rate, and even the district with the lowest birth rate is having 20% higher BR compared to Germany. In 2008, the birth rate in Perm Krai was 8% higher than that of 2007. Close to 35.5 thousand births were recorded with the heaviest increases in City of Perm (+11%) and Komi-Permyak Autonomous Okrug (+18%). Among the districts, Kudymkar City recorded a 46% rise in birth rates for 2008 compared to 2007, while Usolsky recorded a 31% rise and Kyshertsky recorded a 29% rise. In 13 of the districts, there were more births than deaths, among them Ordynsky, Karagaysky, Kudimkar, Chernushynsky, Chaykovsky & Permsky.[18]

Total fertility rate:

1995 - 1,31 | 2000 - 1,26 | 2005 - 1,34 | 2006 - 1,34 | 2007 - 1,47 | 2008 - 1,57 | 2009 - 1,61 | 2010 - 1,77 |[19] 2011 - 1,78 |[20] 2012 - 1,90(e)

Demographics for 2007

[21]

District	Population	Births	Deaths	BR	DR	NGR
Perm Krai	**2,718,227**	**32,747**	**42,680**	**12.05**	**15.70**	**-0.37%**
Bolshesosnovsky	14,292	219	268	15.32	18.75	-0.34%
Vereshchagynsky	43,410	710	744	16.36	17.14	-0.08%
Gornozavodsky	27,885	396	582	14.20	20.87	-0.65%
Yelovsky	12,299	176	229	14.31	18.62	-0.43%
Kochevsky	12,356	181	196	14.65	15.86	-0.12%
Ilyinsky	20,311	278	408	13.69	20.09	-0.64%
Karagaysky	23,907	360	381	15.06	15.94	-0.08%
Kishertsky	14,578	158	280	10.84	19.21	-0.83%
Krasnovishersky	26,287	335	455	12.74	17.31	-0.45%
Kuyedinsky	30,576	446	489	14.59	15.99	-0.14%
Kungursky	46,370	727	720	15.68	15.53	0.01%
Kudymkarsky	27,922	429	676	15.36	24.21	-0.88%
Nitvensky	45,552	656	858	14.40	18.84	-0.44%
Oktyabrsky	34,789	481	585	13.83	16.82	-0.30%
Ordinsky	16,185	247	241	15.26	14.89	0.04%
Osinsky	32,074	410	550	12.78	17.15	-0.44%
Okhansky	17,180	202	359	11.76	20.90	-0.91%
Ochyorsky	24,651	366	455	14.85	18.46	-0.36%
Permsky	87,342	1,275	1,383	14.60	15.83	-0.12%
Sivinsky	16,797	297	266	17.68	15.84	0.18%
Solikamsky	17,637	203	265	11.51	15.03	-0.35%
Suksunsky	20,925	294	349	14.05	16.68	-0.26%
Uynsky	12,631	180	212	14.25	16.78	-0.25%
Usolsky	13,788	176	299	12.76	21.69	-0.88%
Chastinsky	14,450	207	254	14.33	17.58	-0.33%
Cherdynsky	32,522	342	542	10.52	16.67	-0.61%
Gaynsky	16,106	214	268	13.29	16.64	-0.33%
Chernushinsky	52,231	811	638	15.53	12.21	0.33%
Perm	987,246	10,094	13,167	10.22	13.34	-0.31%
Alexandrovsky	34,554	427	662	12.36	19.16	-0.68%
Berezniki	165,950	1,822	2,484	10.98	14.97	-0.40%
Gremyachinsky	15,075	178	367	11.81	24.34	-1.25%
Gubakhinsky	40,086	448	826	11.18	20.61	-0.92%
Dobryansky	61,365	792	950	12.91	15.48	-0.26%
Kizelovsky	30,837	381	827	12.36	26.82	-1.43%

Krasnokamsky	40,393	816	1,264	20.20	31.29	-0.64%
Kungur	68,074	872	981	12.81	14.41	-0.16%
Lysvensky	82,921	1,016	1,598	12.25	19.27	-0.70%
Solikamsk	97,269	1,165	1,464	11.98	15.05	-0.30%
Chaykovsky	108,617	1,374	1,332	12.65	12.26	0.04%
Chusovskoy	73,314	879	1,531	11.99	20.88	-0.89%
Yurlinsky	11,046	200	253	18.11	22.90	-0.48%
Kosinsky	7,716	130	160	16.85	20.74	-0.38%
Yusvinsky	22,626	288	475	12.73	20.99	-0.82%
Kudymkar	30,964	385	511	12.43	16.50	-0.40%
Bardymsky	27,529	361	537	13.11	19.51	-0.64%
Beryozovsky	17,901	230	281	12.85	15.70	-0.28%

Religion

Religion in Perm Krai (2012)[22] [23]

Russian Orthodox (43%)

Unaffiliated Christian (5%)

Islam (4%)

Rodnovery (2%)

Other Orthodox (1%)

Starovery (1%)

Spiritual but not religious (24%)

Atheist and non-religious (14%)

Other and undeclared (8%)

As of a 2012 official survey[22] 43% of the population of Perm Krai adheres to the Russian Orthodox Church, 5% declares to be generically unaffiliated Christian (excluding Catholic and Protestant), 4% are Muslims, 2% are Rodnovers, 1% are Starovers, 1% follows other Orthodox Churches, 8% follows other religion or did not give an answer to the survey. In addition, 24% of the population declares to be "spiritual but not religious" and 14% to be atheist.[22]

Major attractions

Perm Krai is home of several museums:

- Perm State Art Galery
- Perm Museum of Local History
- Museum of Motovilikha Plants
- Architectural-Ethnographic Museum Khokhlovka *and others.*

Perm State Art Gallery

Numerous architectural monuments are located in the small town of Usolye, in north of Perm Krai. Particularly important are the Saviour Cathedral, with a separate bell tower, and House of Stroganov.

There are many theaters in Perm, including the Perm Opera and Ballet Theater, the Perm Academic Theater, the Puppet Theater, the Theater for Young Spectators, the Theater "Near Bridge", and others.

Khokhlovka

There are many temples and convents in Perm Krai. The most significant of them are: Belogorsky Convent located in 85 km from Perm, Sludskaya Church, Fedosievskaya Church, Perm Mosque and others.

Ordinsky is home to the Orda underwater caves. http://ordacave.ru/en/ [24] Located near Orda village in Perm region, Ural, Orda Cave is also the biggest underwater gypsum crystal cave in the world.

References

[1] Президент Российской Федерации. Указ №849 от 13 мая 2000 г. «О полномочном представителе Президента Российской Федерации в федеральном округе». Вступил в силу 13 мая 2000 г. Опубликован: "Собрание законодательства РФ", №20, ст. 2112, 15 мая 2000 г. (President of the Russian Federation. Decree #849 of May 13, 2000 *On the Plenipotentiary Representative of the President of the Russian Federation in a Federal District*. Effective as of May 13, 2000.).

[2] Госстандарт Российской Федерации. №ОК 024-95 27 декабря 1995 г. «Общероссийский классификатор экономических регионов. 2. Экономические районы», в ред. Изменения №5/2001 ОКЭР. (Gosstandart of the Russian Federation. #OK 024-95 December 27, 1995 *Russian Classification of Economic Regions. 2. Economic Regions*, as amended by the Amendment #5/2001 OKER.).

[3] Федеральный конституционный закон №1-ФКЗ от 25 марта 2004 г «Об образовании в составе Российской Федерации нового субъекта Российской Федерации в результате объединения Пермской области и Коми-Пермяцкого автономного округа», в ред. федерального конституционного закона №1-ФКЗ от 12 апреля 2006 г. (Federal Constitutional Law #1-FKZ of March 25, 2004 *On Establishment Within the Russian Federation of a New Federal Subject of the Russian Federation as a Result of the Merger of Perm Oblast and Komi-Permyak Autonomous Okrug*, as amended by the Federal Constitutional Law #1-FKZ of April 12, 2006.).

[4] Charter, Article 15.3

[5] Official website of Perm Krai. Viktor Fyodorovich Basargin, Governor of Perm Krai (http://www.perm.ru/power/gubernator/) (Russian)

[6] Charter, Article 15.4

[7] The density value was calculated by dividing the population reported by the 2010 Census by the area shown in the "Area" field. Please note that this value may not be accurate as the area specified in the infobox is not necessarily reported for the same year as the population.

[8] Правительство Российской Федерации. Постановление №725 от 31 августа 2011 г. « О составе территорий, образующих каждую часовую зону, и порядке исчисления времени в часовых зонах, а также о признании утратившими силу отдельных Постановлений Правительства Российской Федерации (http://www.consultant.ru/online/base/?req=doc;base=LAW;n=118896)». Вступил в силу по истечении 7 дней после дня официального опубликования. Опубликован: "Российская Газета", №197, 6 сентября 2011 г. (Government of the Russian Federation. Resolution #725 of August 31, 2011 *On the Composition of the Territories Included into Each Time Zone and on the Procedures of Timekeeping in the Time Zones, as Well as on Abrogation of Several Resolutions of the Government of the Russian Federation* (http://translate.google.com/translate?hl=en&sl=ru&tl=en&u=http://www.consultant.ru/online/base/?req=doc;base=LAW;n=118896). Effective as of after 7 days following the day of the official publication.).

[9] Official the whole territory of Russia according to Article 68.1 of the Constitution of Russia.

[10] http://www.perm.ru

[11] Полезные ископаемые (http://www.perm.ru/region/geography/minerals/)

[12] http://www.perepis-2010.ru/news/detail.php?ID=6936

[13] Статья о башкирах на сайте РОО "Хомай" (http://www.homay.ru/traditions/4/)

[14] ТАТАРЫ. Энциклопедия «Пермский край» (http://enc.permkultura.ru/showObject.do?object=1803792510)
[15] http://www.gks.ru/free_doc/2012/demo/edn12-12.htm
[16] http://www.gks.ru/bgd/regl/b12_111/Main.htm
[17] http://www.gks.ru/wps/wcm/connect/rosstat_main/rosstat/ru/statistics/publications/catalog/doc_1137674209312
[18] http://www.regnum.ru/news/1114845.html
[19] http://permstat.gks.ru/wps/wcm/connect/rosstat_ts/permstat/ru/statistics/population/
[20] http://www.gks.ru/bgd/regl/b12_50/Main.htm
[21] http://dmr.perm.ru/?/katalog.dirx2x79x143.205*xls
[22] Arena - Atlas of Religions and Nationalities in Russia (http://sreda.org/arena). Sreda.org
[23] 2012 Survey Maps (http://c2.kommersant.ru/ISSUES.PHOTO/OGONIOK/2012/034/ogcyhjk2.jpg). "Ogonek", № 34 (5243), 27/08/2012. *Retrieved 24-09-2012.*
[24] http://ordacave.ru/en/

Sources

- Законодательное Собрание Пермского края. №32-ПК 27 апреля 2007 г. «Устав Пермского края». (Legislative Assembly of Perm Krai. #32-PK April 27, 2007 *Charter of Perm Krai.*).

External links

- Official website of Perm Krai (http://perm.ru/eng/) **(Russian)**
- Perm Chamber of Commerce and Industry (http://invest.permtpp.ru/en/main/)
- Perm State University (http://www.psu.ru/?a=0&lang=Eng&m=008)
- Tourist Resources of Perm Krai (http://visitperm.com)

Russia

Russian Federation

Anthem:
"Государственный гимн Российской Федерации"
"*Gosudarstvenny gimn Rossiyskoy Federatsii*"(transliteration)
"State Anthem of the Russian Federation"

Capital and largest city		Moscow
Official languages		Russian official throughout the country; 27 other languages co-official in various regions
Demonym		Russian
Government		Federal semi-presidential constitutional republic
-	President	Vladimir Putin
-	Prime Minister	Dmitry Medvedev
Legislature		Federal Assembly
-	Upper house	Federation Council
-	Lower house	State Duma
Formation		
-	Kievan Rus'	882
-	Grand Duchy of Moscow	1283
-	Tsardom of Russia	16 January 1547
-	Russian Empire	22 October 1721
-	Russian SFSR	6 November 1917
-	Part of Soviet Union	10 December 1922

-	Russian Federation	25 December 1991
Area		
-	Total	17,098,242 km^2 (1st) 6,592,800 sq mi
-	Water (%)	13[] (including swamps)
Population		
-	2013 estimate	143,400,000[1] (9th)
-	Density	8.4/km^2 (217th) 21.5/sq mi
GDP (PPP)		2012 estimate
-	Total	$ 3.380 Trillion [2]) (PPP; 5th (6th)
-	Per capita	$23,549[3] (43th)
GDP (nominal)		2012 estimate
-	Total	$2.014 trillion[4]) (8th)
-	Per capita	$14,037[5] (50th)
Gini (2008)		42.3[6] medium · 83rd
HDI (2013)		▲ 0.788[] high · 55th
Currency		Russian ruble (RUB)
Time zone		(UTC+3 to +12[a])
Date format		dd.mm.yyyy
Drives on the		right
Calling code		+7
ISO 3166 code		RU
a.	Excluding +5.	

Russia [i]/ˈrʌʃə/ or /ˈrʊʃə/ (Russian: Россия, tr. *Rossiya*, IPA: [rɐˈsʲijə] (listen)), also officially known as the **Russian Federation**[7] (Russian: Российская Федерация, tr. *Rossiyskaya Federatsiya*, IPA: [rɐˈsʲijskəjə fʲɪdʲɪˈrat͡sɨjə] (listen)), is a country in northern Eurasia.[] It is a federal semi-presidential republic, comprising 83 federal subjects. From northwest to southeast, Russia shares land borders with Norway, Finland, Estonia, Latvia, Lithuania and Poland (both with Kaliningrad Oblast), Belarus, Ukraine, Georgia, Azerbaijan, Kazakhstan, China, Mongolia, and North Korea. It shares maritime borders with Japan by the Sea of Okhotsk, and the US state of Alaska across the Bering Strait. At 17075400 square kilometres (sq mi), Russia is the largest country in the world, covering more than one-eighth of the Earth's inhabited land area. Russia is also the world's ninth most populous nation with 143 million people as of 2012.[8] Extending across the whole of northern Asia and much of Europe, Russia spans nine time zones and incorporates a wide range of environments and landforms.

The nation's history began with that of the East Slavs, who emerged as a recognizable group in Europe between the 3rd and 8th centuries AD.[] Founded and ruled by a Varangian warrior elite and their descendants, the medieval state of Rus arose in the 9th century. In 988 it adopted Orthodox Christianity from the Byzantine Empire,[] beginning the synthesis of Byzantine and Slavic cultures that defined Russian culture for the next millennium.[] Rus' ultimately disintegrated into a number of smaller states; most of the Rus' lands were overrun by the Mongol invasion and became tributaries of the nomadic Golden Horde.[9] The Grand Duchy of Moscow gradually reunified the surrounding Russian principalities, achieved independence from the Golden Horde, and came to dominate the cultural and political legacy of Kievan Rus'. By the 18th century, the nation had greatly expanded through conquest, annexation, and exploration to become the Russian Empire, which was the third largest empire in history, stretching from Poland in Europe to Alaska in North America.[10] [11]

Following the Russian Revolution, the Russian Soviet Federative Socialist Republic became the largest and leading constituent of the Soviet Union, the world's first constitutionally socialist state and a recognized superpower,[12] which played a decisive role in the Allied victory in World War II.[13] [14] The Soviet era saw some of the most significant technological achievements of the 20th century, including the world's first spacecraft, and the first astronaut. The Russian Federation became the successor state of the Russian SFSR following the dissolution of the Soviet Union in 1991, and is recognized as the continuing legal personality of the All-Union state.[]

The Russian economy ranking as the eighth largest by nominal GDP and sixth largest by purchasing power parity. Russia's extensive mineral and energy resources, the largest reserves in the world,[15] have made it one of the largest producers of oil and natural gas globally.[16] [17] The country is one of the five recognized nuclear weapons states and possesses the largest stockpile of weapons of mass destruction.[] Russia is a great power and a permanent member of the United Nations Security Council, a member of the G8, G20, the Council of Europe, the Asia-Pacific Economic Cooperation, the Shanghai Cooperation Organisation, the Eurasian Economic Community, the Organisation for Security and Cooperation in Europe (OSCE), the World Trade Organisation (WTO), and is the leading member of the Commonwealth of Independent States.

Etymology

The name *Russia* is derived from Rus, a medieval state populated mostly by the East Slavs. However, this proper name became more prominent in the later history, and the country typically was called by its inhabitants "Русская Земля" (russkaya zemlya) which could be translated as "Russian Land" or "Land of Rus'". In order to distinguish this state from other states derived from it, it is denoted as *Kievan Rus'* by modern historiography. The name *Rus* itself comes from Rus people, a group of Varangians (possibly Swedish Vikings)[18] [19] who founded the state of Rus (Русь).

An old Latin version of the name Rus' was Ruthenia, mostly applied to the western and southern regions of Rus' that were adjacent to Catholic Europe. The current name of the country, Россия (Rossiya), comes from the Byzantine Greek designation of the Kievan Rus', Ρωσσία *Rossía*—spelt Ρωσία (*Rosía* pronounced [ro'sia]) in Modern Greek.[20]

History

Early periods

Further information: Eurasian nomads, Scythia, Bosporan Kingdom, Goths, Khazars, and East Slavs

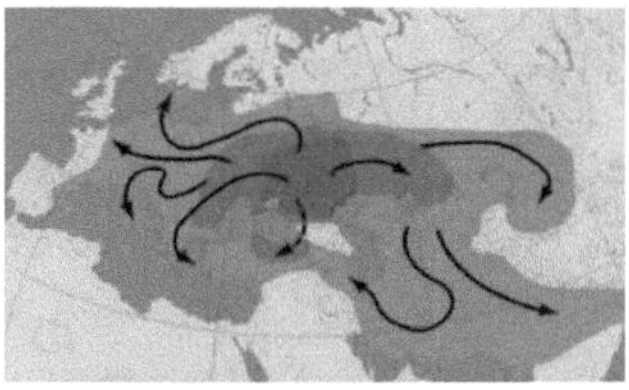
Kurgan hypothesis: South Russia as the urheimat of Indo-European peoples

One of the first modern human bones, dated to 35,000 years old, was found in Russia, in Kostenki on the Don River banks. The only remains of the Denisova hominin that lived about 41,000 years ago were discovered in Denisova Cave (South Siberia).

In prehistoric times the vast steppes of Southern Russia were home to tribes of nomadic pastoralists.[] Remnants of these steppe civilizations were discovered in such places as Ipatovo,[] Sintashta,[21] Arkaim,[22] and Pazyryk,[23] which bear the earliest known traces of mounted warfare, a key feature in the nomadic way of life.

In classical antiquity, the Pontic Steppe was known as Scythia. Since the 8th century BC, Ancient Greek traders brought their civilization to the trade emporiums in Tanais and Phanagoria. The Romans settled on the western part of the Caspian Sea, where their empire stretched towards the east.[24] In 3rd – 4th centuries AD a semi-legendary Gothic kingdom of Oium existed in Southern Russia till it was overrun by Huns. Between the 3rd and 6th centuries AD, the Bosporan Kingdom, a Hellenistic polity which succeeded the Greek colonies,[25] was also overwhelmed by nomadic invasions led by warlike tribes, such as the Huns and Eurasian Avars.[26] A Turkic people, the Khazars, ruled the lower Volga basin steppes between the Caspian and Black Seas until the 10th century.[27]

The ancestors of modern Russians are the Slavic tribes, whose original home is thought by some scholars to have been the wooded areas of the Pinsk Marshes.[28] The East Slavs gradually settled Western Russia in two waves: one moving from Kiev toward present-day Suzdal and Murom and another from Polotsk toward Novgorod and Rostov. From the 7th century onwards, the East Slavs constituted the bulk of the population in Western Russia[29] and slowly but peacefully assimilated the native Finno-Ugric peoples, including the Merya, the Muromians, and the Meshchera.

Kievan Rus

The establishment of the first East Slavic states in the 9th century coincided with the arrival of *Varangians*, the traders, warriors and settlers from the Baltic Sea region. Primarily they were Vikings of Scandinavian origin, who ventured along the waterways extending from the eastern Baltic to the Black and Caspian Seas.[30] According to the Primary Chronicle, a Varangian from Rus' people, named Rurik, was elected ruler of Novgorod in 862. In 882 his successor Oleg ventured south and conquered Kiev,[31] which had been previously paying tribute to the Khazars, founding Kievan Rus'. Oleg, Rurik's son Igor and Igor's son Sviatoslav subsequently subdued all local East Slavic tribes to Kievan rule, destroyed the Khazar khaganate and launched several military expeditions to Byzantium and Persia.

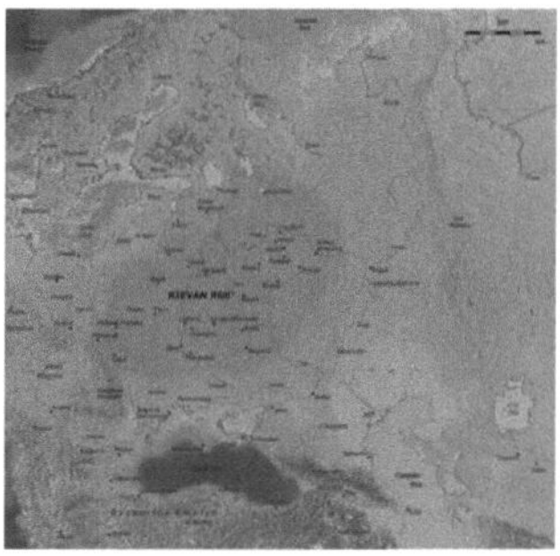
Kievan Rus' in the 11th century

In the 10th to 11th centuries Kievan Rus' became one of the largest and most prosperous states in Europe.[32] The reigns of Vladimir the Great (980–1015) and his son Yaroslav the Wise (1019–1054) constitute the Golden Age of Kiev, which saw the acceptance of Orthodox Christianity from Byzantium and the creation of the first East Slavic written legal code, the *Russkaya Pravda*.

In the 11th and 12th centuries, constant incursions by nomadic Turkic tribes, such as the Kipchaks and the Pechenegs, caused a massive migration of Slavic populations to the safer, heavily forested regions of the north, particularly to the area known as Zalesye.[33]

The Baptism of Kievans, by Klavdy Lebedev

The age of feudalism and decentralization was marked by constant in-fighting between members of the Rurik Dynasty that ruled Kievan Rus' collectively. Kiev's dominance waned, to the benefit of Vladimir-Suzdal in the north-east, Novgorod Republic in the north-west and Galicia-Volhynia in the south-west.

Ultimately Kievan Rus' disintegrated, with the final blow being the Mongol invasion of 1237–40,[34] that resulted in the destruction of Kiev[35] and the death of about half the population of Rus'.[36] The invading Mongol elite, together with their conquered Turkic subjects (Cumans, Kipchaks, Bulgars) became known as Tatars, formed the state of the Golden Horde, which pillaged the Russian principalities; the Mongols ruled the Cuman-Kipchak confederation and Volga Bulgaria (modern-day southern and central expanses of Russia) for over two centuries.[37]

Galicia-Volhynia was eventually assimilated by the Polish-Lithuanian Commonwealth, while the Mongol-dominated Vladimir-Suzdal and Novgorod Republic, two regions on the periphery of Kiev, established the basis for the modern Russian nation.[] The Novgorod together with Pskov retained some degree of autonomy during the time of the Mongol yoke and were largely spared the atrocities that affected the rest of the country. Led by Prince Alexander Nevsky, Novgorodians repelled the invading Swedes in the Battle of the Neva in 1240, as well as the Germanic crusaders in the Battle of the Ice in 1242, breaking their attempts to colonize the Northern Rus'.

Grand Duchy of Moscow

Sergius of Radonezh blessing Dmitry Donskoy in Trinity Sergius Lavra, before the Battle of Kulikovo, depicted in a painting by Ernst Lissner

The most powerful successor state to Kievan Rus' was the Grand Duchy of Moscow ("Moscovy" in the Western chronicles), initially a part of Vladimir-Suzdal. While still under the domain of the Mongol-Tatars and with their connivance, Moscow began to assert its influence in the Central Rus' in the early 14th century, gradually becoming the main leading force in the process of the Rus' lands' reunification and expansion of Russia.

Those were hard times, with frequent Mongol-Tatar raids and agriculture suffering from the beginning of the Little Ice Age. As in the rest of Europe, plagues hit Russia somewhere once every five or six years from 1350 to 1490. However, because of the lower population density and better hygiene (widespread practicing of banya, the wet steam bath),[38] the population loss caused by plagues was not so severe as in the Western Europe, and the pre-Plague populations were reached in Russia as early as 1500.[39]

Led by Prince Dmitry Donskoy of Moscow and helped by the Russian Orthodox Church, the united army of Russian principalities inflicted a milestone defeat on the Mongol-Tatars in the Battle of Kulikovo in 1380. Moscow gradually absorbed the surrounding principalities, including the formerly strong rivals, such as Tver and Novgorod.

Ivan III (*the Great*) finally threw off the control of the Golden Horde, consolidated the whole of Central and Northern Rus' under Moscow's dominion, and was the first to take the title "Grand Duke of all the Russias".[40] After the fall of Constantinople in 1453, Moscow claimed succession to the legacy of the Eastern Roman Empire. Ivan III married Sophia Palaiologina, the niece of the last Byzantine emperor Constantine XI, and made the Byzantine double-headed eagle his own, and eventually Russian, coat-of-arms.

Tsardom of Russia

Tsar Ivan the Terrible by Victor Vasnetsov

In development of the Third Rome ideas, the Grand Duke Ivan IV (the "Terrible"[41]) was officially crowned the first Tsar ("Caesar") of Russia in 1547. The Tsar promulgated a new code of laws (Sudebnik of 1550), established the first Russian feudal representative body (Zemsky Sobor) and introduced local self-management into the rural regions.[42] [43]

During his long reign, Ivan the Terrible nearly doubled the already large Russian territory by annexing the three Tatar khanates (parts of disintegrated Golden Horde): Kazan and Astrakhan along the Volga River, and Sibirean Khanate in South Western Siberia. Thus by the end of the 16th century Russia was transformed into a multiethnic, multidenominational and transcontinental state.

However, the Tsardom was weakened by the long and unsuccessful Livonian War against the coalition of Poland, Lithuania, and Sweden for access to the Baltic coast and sea trade.[44] At the same time the Tatars of the Crimean Khanate, the only remaining successor to the Golden Horde, continued to raid Southern Russia.[45] In effort to restore the Volga khanates, Crimeans and their Ottoman allies invaded central Russia and were even able to burn down parts of Moscow in 1571.[46] But next year the large invading army was thoroughly defeated by Russians in the Battle of Molodi, forever eliminating the threat of the Ottoman-Crimean expansion into Russia. The raids of Crimeans, however, didn't cease until the late 17th century, though the construction of new fortification lines across Southern Russia, such as the Great Abatis Line, constantly narrowed the area accessible to incursions.

Monument to Minin and Pozharsky in Moscow

The death of Ivan's sons marked the end of the ancient Rurik Dynasty in 1598, and in combination with the famine of 1601–03[47] led to the civil war, the rule of pretenders and foreign intervention during the Time of Troubles in the early 17th century.[48] Polish-Lithuanian Commonwealth occupied parts of Russia, including Moscow. In 1612, the Poles were forced to retreat by the Russian volunteer corps, led by two national heroes, merchant Kuzma Minin and Prince Dmitry Pozharsky. The Romanov Dynasty acceded the throne in 1613 by the decision of Zemsky Sobor, and the country started its gradual recovery from the crisis.

Russia continued its territorial growth through the 17th century, which was the age of Cossacks. Cossacks were warriors organized into military communities, resembling pirates and pioneers of the New World. In 1648, the peasants of Ukraine joined the Zaporozhian Cossacks in rebellion against Poland-Lithuania during the Khmelnytsky Uprising, because of the social and religious oppression they suffered under Polish rule. In 1654, the Ukrainian leader, Bohdan Khmelnytsky, offered to place Ukraine under the protection of the Russian Tsar, Aleksey I. Aleksey's acceptance of this offer led to another Russo-Polish War (1654–1667). Finally, Ukraine was split along the Dnieper River, leaving the western part (or Right-bank Ukraine) under Polish rule and eastern part (Left-bank Ukraine and Kiev) under Russian. Later, in 1670–71 the Don Cossacks led by Stenka Razin initiated a major uprising in the Volga Region, but the Tsar's troops were successful in defeating the rebels.

In the east, the rapid Russian exploration and colonisation of the huge territories of Siberia was led mostly by Cossacks hunting for valuable furs and ivory. Russian explorers pushed eastward primarily along the Siberian River Routes, and by the mid-17th century there were Russian settlements in Eastern Siberia, on the Chukchi Peninsula, along the Amur River, and on the Pacific coast. In 1648, the Bering Strait between Asia and North America was passed for the first time by Fedot Popov and Semyon Dezhnyov.

Imperial Russia

Peter the Great, the first Emperor of Russia

Under Peter the Great, Russia was proclaimed an Empire in 1721 and became recognized as a world power. Ruling from 1682 to 1725, Peter defeated Sweden in the Great Northern War, forcing it to cede West Karelia and Ingria (two regions lost by Russia in the Time of Troubles),[49] as well as Estland and Livland, securing Russia's access to the sea and sea trade.[50] On the Baltic Sea Peter founded a new capital called Saint Petersburg, later known as Russia's *Window to Europe*. Peter the Great's reforms brought considerable Western European cultural influences to Russia.

The reign of Peter I's daughter Elizabeth in 1741–62 saw Russia's participation in the Seven Years' War (1756–63). During this conflict Russia annexed East Prussia for a while and even took Berlin. However, upon Elisabeth's death, all these conquests were returned to Kingdom of Prussia by pro-Prussian Peter III of Russia.

Catherine II (*the Great*), who ruled in 1762–96, presided over the Age of Russian Enlightenment. She extended Russian political control over the Polish-Lithuanian Commonwealth and incorporated most of its territories into Russia during the Partitions of Poland, pushing the Russian frontier westward into Central Europe. In the south, after successful Russo-Turkish Wars against the Ottoman Empire, Catherine advanced Russia's boundary to the Black Sea, defeating the Crimean Khanate. As a result of victories over the Ottomans, by the early 19th century Russia also made significant territorial gains in Transcaucasia. This continued with Alexander I's (1801–25) wresting of Finland from the weakened kingdom of Sweden in 1809 and of Bessarabia from the Ottomans in 1812. At the same time Russians colonized Alaska and even founded settlements in California, like Fort Ross.

In 1803–06, the first Russian circumnavigation was made, later followed by other notable Russian sea exploration voyages. In 1820 a Russian expedition discovered the continent of Antarctica.

The Russian Empire in 1866 and its spheres of influence

In alliances with various European countries, Russia fought against Napoleon's France. The French invasion of Russia at the height of Napoleon's power in 1812 failed miserably as the obstinate resistance in combination with the bitterly cold Russian Winter led to a disastrous defeat of invaders, in which more than 95% of the pan-European Grande Armée perished.[51] Led by Mikhail Kutuzov and Barclay de Tolly, the Russian army ousted Napoleon from the country and drove through Europe in the war of the Sixth Coalition, finally entering Paris. Alexander I headed Russia's delegation at the Congress of Vienna that defined the map of post-Napoleonic Europe.

The officers of the Napoleonic Wars brought ideas of liberalism back to Russia with them and attempted to curtail the tsar's powers during the abortive Decembrist revolt of 1825. At the end of the conservative reign of Nicolas I (1825–55), a zenith period of Russia's power and influence in Europe was disrupted by defeat in the Crimean War. Between 1847 and 1851, a massive wave of Asiatic cholera swept over Russia, claiming about one million lives.[52]

Nicholas's successor Alexander II (1855–81) enacted significant changes in the country, including the emancipation reform of 1861. These *Great Reforms* spurred industrialization and modernized the Russian army, which had successfully liberated Bulgaria from Ottoman rule in 1877–78 Russo-Turkish War.

Vladimir Lenin, the leader of the Great October Socialist Revolution

The late 19th century saw the rise of various socialist movements in Russia. Alexander II was killed in 1881 by revolutionary terrorists, and the reign of his son Alexander III (1881–94) was less liberal but more peaceful. The last Russian Emperor, Nicholas II (1894–1917), was unable to prevent the events of the Russian Revolution of 1905, triggered by the unsuccessful Russo-Japanese War and the demonstration incident known as Bloody Sunday. The uprising was put down, but the government was forced to concede major reforms, including granting the freedoms of speech and assembly, the legalization of political parties, and the creation of an elected legislative body, the State Duma of the Russian Empire. Migration to Siberia increased rapidly in the early 20th century, particularly during the Stolypin agrarian reform. Between 1906 and 1914 more than four million settlers arrived in that region.[53]

In 1914, Russia entered World War I in response to Austria-Hungary's declaration of war on Russia's ally Serbia, and fought across multiple fronts while isolated from its Triple Entente allies. In 1916, the Brusilov Offensive of the Russian Army almost completely destroyed the military of Austria-Hungary. However, the already-existing public distrust of the regime was deepened by the rising costs of war, high casualties, and rumors of corruption and treason. All this formed the climate for the Russian Revolution of 1917, carried out in two major acts.

Revolution and Russian Republic

The February Revolution forced Nicholas II to abdicate; he and his family were imprisoned and later executed during the Russian Civil War. The monarchy was replaced by a shaky coalition of political parties that declared itself the Provisional Government. An alternative socialist establishment existed alongside, the Petrograd Soviet, wielding power through the democratically elected councils of workers and peasants, called *Soviets*. The rule of the new authorities only aggravated the crisis in the country, instead of resolving it. Eventually, the October Revolution, led by Bolshevik leader Vladimir Lenin, overthrew the Provisional Government and gave full governing power to the Soviets, leading to the creation of the world's first socialist state.

Soviet Russia and civil war

Following the October Revolution, a civil war broke out between the anti-communist White movement and the new Soviet regime with its Red Army. Bolshevist Russia lost its Ukrainian, Polish, Baltic, and Finnish territories by signing the Treaty of Brest-Litovsk that concluded hostilities with the Central Powers of World War I. The Allied powers launched an unsuccessful military intervention in support of anti-Communist forces. In the meantime both the Bolsheviks and White movement carried out campaigns of deportations and executions against each other, known respectively as the Red Terror and White Terror. By the end of the civil war, the Russian economy and its infrastructure were heavily damaged. Millions became White émigrés,[54] and the Povolzhye famine of 1921 claimed up to 5 million victims.[55]

The symbols of the early Soviet era: Tatlin's Tower project and the giant *Worker and Kolkhoz Woman* sculpture group

Soviet Union

The Russian Soviet Federative Socialist Republic (called *Russian Socialist Federative Soviet Republic* at the time) together with the Ukrainian, Byelorussian, and Transcaucasion Soviet Socialist Republics, formed the Union of Soviet Socialist Republics (USSR), or Soviet Union, on 30 December 1922. Out of the 15 republics that would make up the USSR, the largest in size and over half of the total USSR population was the Russian SFSR, which came to dominate the union for its entire 69-year history.

Following Lenin's death in 1924, a troika was designated to govern the Soviet Union. However, Joseph Stalin, an elected General Secretary of the Communist Party, managed to suppress all opposition groups within the party and consolidate power in his hands. Leon Trotsky, the main proponent of world revolution, was exiled from the Soviet Union in 1929, and Stalin's idea of Socialism in One Country became the primary line. The continued internal struggle in the Bolshevik party culminated in the Great Purge, a period of mass repressions in 1937–38, during which hundreds of thousands of people were executed, including original party members and military leaders accused of coup d'état plots.[56]

The Red Army is greeted in Bucharest, August 1944

Under Stalin's leadership, the government launched a planned economy, industrialisation of the largely rural country, and collectivization of its agriculture. During this period of rapid economic and social change, millions of people were sent to penal labor camps,[57] including many political convicts for their opposition to Stalin's rule; millions were deported and exiled to remote areas of the Soviet Union.[57] The transitional disorganisation of the country's agriculture, combined with the harsh state policies and a drought, led to the Soviet famine of 1932–1933.[58] The Soviet Union, though with a heavy price, was transformed from a largely agrarian economy to a major industrial powerhouse in a short span of time.

The Appeasement policy of Great Britain and France towards Adolf Hitler's annexation of Austria and Czechoslovakia enlarged the might of Nazi Germany and put a threat of war to the Soviet Union. Around the same time the Third Reich allied with the Empire of Japan, a rival of the USSR in the Far East and an open enemy of the USSR in the Soviet–Japanese Border Wars in 1938–39.

In August 1939, after another failure of attempts to establish an anti-Nazi alliance with Britain and France, the Soviet government decided to improve relations with Germany by concluding the Molotov-Ribbentrop Pact, pledging non-aggression between the two countries and dividing their spheres of influence in Eastern Europe. While Hitler conquered Poland, France and other countries actied on single front at the start of World War II, the USSR was able to build up its military and claim some of the former territories of the Russian Empire as a result of the Soviet invasion of Poland, Winter War and the occupation of the Baltic states.

On 22 June 1941, Nazi Germany broke the non-aggression treaty and invaded the Soviet Union with the largest and most powerful invasion force in human history,[59] opening the largest theater of World War II. Although the German army had considerable early success, their attack was halted in the Battle of Moscow. Subsequently the Germans were dealt major defeats first at the Battle of Stalingrad in the winter of 1942–43,[60] and then in the Battle of Kursk in the summer of 1943. Another German failure was the Siege of Leningrad, in which the city was fully blockaded on land between 1941–44 by German and Finnish forces, and suffered starvation and more than a million deaths, but never surrendered.[61] Under Stalin's administration and the leadership of such commanders as Georgy Zhukov and Konstantin Rokossovsky, Soviet forces took Eastern Europe in 1944–45 and captured Berlin in May 1945. In August 1945 the Soviet Army ousted the Japanese from China's Manchukuo and North Korea, contributing to the allied victory over Japan.

First human to travel into space, Yuri Gagarin

The 1941–45 period of World War II is known in Russia as the *Great Patriotic War*. During this conflict, which included many of the most lethal battle operations in human history, Soviet military and civilian deaths were 10.6 million and 15.9 million respectively,[62] accounting for about a third of all World War II casualties. The full demographic loss to the Soviet peoples was even greater.[63] The Soviet economy and infrastructure suffered massive devastation[64] but the Soviet Union emerged as an acknowledged military superpower on the continent.

The Red Army occupied Eastern Europe after the war, including East Germany. Dependent socialist governments were installed in the Eastern Bloc satellite states. Becoming the world's second nuclear weapons power, the USSR established the Warsaw Pact alliance and entered into a struggle for global dominance, known as the Cold War, with the United States and NATO. The Soviet Union supported revolutionary movements across the world, including the newly formed People's Republic of China, the Democratic People's Republic of Korea and, later on, the Republic of Cuba. Significant amounts of the Soviet resources were allocated in aid to the other socialist states.[65]

After Stalin's death and a short period of collective rule, new leader Nikita Khrushchev denounced the cult of personality of Stalin and launched the policy of de-Stalinization. The penal labor system was reformed and many prisoners were released and rehabilitated (many of them posthumously).[66] The general easement of repressive policies became known later as the Khrushchev Thaw. At the same time, tensions with the United States heightened when the two rivals clashed over the deployment of the U.S. Jupiter missiles in Turkey and Soviet missiles in Cuba.

In 1957, the Soviet Union launched the world's first artificial satellite, *Sputnik 1*, thus starting the Space Age. Russian cosmonaut Yuri Gagarin became the first human to orbit the Earth aboard *Vostok 1* manned spacecraft on 12 April 1961.

Sputnik 1 was the world's first artificial satellite

Following the ousting of Khrushchev in 1964, another period of collective rule ensued, until Leonid Brezhnev became the leader. The era of 1970s and the early 1980s was designated later as the Era of Stagnation, a period when the economic growth slowed and social policies became static. The 1965 Kosygin reform, aimed for partial decentralization of the Soviet economy and shifted the emphasis from heavy industry and weapons to light industry and consumer goods, but was stifled by the conservative Communist leadership.

In 1979, after a Communist-led revolution in Afghanistan, Soviet forces entered that country at request of the new regime. The occupation drained economic resources and dragged on without achieving meaningful political results. Ultimately the Soviet Army was withdrawn from Afghanistan in 1989 due to international opposition, persistent anti-Soviet guerilla warfare, and a lack of support by Soviet citizens.

From 1985 onwards, the last Soviet leader Mikhail Gorbachev, who sought to enact liberal reforms in the Soviet system, introduced the policies of *glasnost* (openness) and *perestroika* (restructuring) in an attempt to end the period of economic stagnation and to democratise the government. This, however, led to the rise of strong nationalist and separatist movements. Prior to 1991, the Soviet economy was the second largest in the world,[67] but during its last years it was afflicted by shortages of goods in grocery stores, huge budget deficits, and explosive growth in money supply leading to inflation.[68]

By 1991, economic and political turmoil began to boil over, as the Baltic republics chose to secede from the Union. On 17 March, a referendum was held, to which the vast majority of participating citizens voted in favour of preserving the Soviet Union as a renewed federation. In August 1991, a coup d'état attempt by members of Gorbachev's government, directed against Gorbachev and aimed at preserving the Soviet Union, instead led to the end of the Communist Party of the Soviet Union. Despite the will expressed by the people, on 25 December 1991, the USSR was dissolved into 15 post-Soviet states.

Russian Federation

Boris Yeltsin was elected the President of Russia in June 1991, in the first direct presidential election in Russian history. During and after the Soviet disintegration, wide-ranging reforms including privatization and market and trade liberalization were undertaken,[] including radical changes along the lines of "shock therapy" as recommended by the United States and International Monetary Fund.[69] All this resulted in a major economic crisis, characterized by 50% decline of both GDP and industrial output between 1990–95.[] [70]

Moscow International Business Center under construction

The privatization largely shifted control of enterprises from state agencies to individuals with inside connections in the government. Many of the newly rich moved billions in cash and assets outside of the country in an enormous capital flight.[71] The depression of the economy led to the collapse of social services; the birth rate plummeted while the death rate skyrocketed. Millions plunged into poverty, from 1.5% level of poverty in the late Soviet era, to 39–49% by mid-1993.[] The 1990s saw extreme corruption and lawlessness, the rise of criminal gangs and violent crime.[72]

The 1990s were plagued by armed conflicts in the North Caucasus, both local ethnic skirmishes and separatist Islamist insurrections. From the time Chechen separatists declared independence in the early 1990s, an intermittent guerrilla war has been fought between the rebel groups and the Russian military. Terrorist attacks against civilians carried out by separatists, most notably the Moscow theater hostage crisis and Beslan school siege, caused hundreds of deaths and drew worldwide attention.

Russia took up the responsibility for settling the USSR's external debts, even though its population made up just half of the population of the USSR at the time of its dissolution.[73] High budget deficits caused the 1998 Russian financial crisis[74] and resulted in a further GDP decline.[]

On 31 December 1999, President Yeltsin resigned, handing the post to the recently appointed Prime Minister, Vladimir Putin, who then won the 2000 presidential election. Putin suppressed the Chechen insurgency, although sporadic violence still occurs throughout the Northern Caucasus. High oil prices and the initially weak currency followed by increasing domestic demand, consumption and investments has helped the economy to grow for nine straight years, improving the standard of living and increasing Russia's influence on the world stage.[17] While many reforms made during the Putin presidency have been generally criticized by Western nations as un-democratic,[75] Putin's leadership over the return of order, stability, and progress has won him widespread popularity in Russia.[76]

On 2 March 2008, Dmitry Medvedev was elected President of Russia, whilst Putin became Prime Minister. Putin returned to the presidency following the 2012 presidential elections, and Medvedev was appointed Prime Minister.

Politics

Governance

Moscow Kremlin, the working residence of the President of Russia

According to the Constitution of Russia, the country is a federation and semi-presidential republic, wherein the President is the head of state[77] and the Prime Minister is the head of government. The Russian Federation is fundamentally structured as a multi-party representative democracy, with the federal government composed of three branches:

- Legislative: The bicameral Federal Assembly of Russia, made up of the 450-member State Duma and the 166-member Federation Council, adopts federal law, declares war, approves treaties, has the power of the purse and the power of impeachment of the President.
- Executive: The President is the Supreme Commander-in-Chief of the Armed Forces, can veto legislative bills before they become law, and appoints the Government of Russia (Cabinet) and other officers, who administer and enforce federal laws and policies.
- Judiciary: The Constitutional Court, Supreme Court, Supreme Court of Arbitration and lower federal courts, whose judges are appointed by the Federation Council on the recommendation of the President, interpret laws and can overturn laws they deem unconstitutional.

The president is elected by popular vote for a six-year term (eligible for a second term, but not for a third consecutive term).[78] Ministries of the government are composed of the Premier and his deputies, ministers, and selected other individuals; all are appointed by the President on the recommendation of the Prime Minister (whereas the appointment of the latter requires the consent of the State Duma). Leading political parties in Russia include United Russia, the Communist Party, the Liberal Democratic Party of Russia, and A Just Russia.

Foreign relations

Leaders of the BRIC nations in 2008: (l-r) Manmohan Singh of India, Dmitry Medvedev of Russia, Hu Jintao of China and Luiz Inácio Lula da Silva of Brazil

The Russian Federation is recognized in international law as a successor state of the former Soviet Union.[] Russia continues to implement the international commitments of the USSR, and has assumed the USSR's permanent seat in the UN Security Council, membership in other international organisations, the rights and obligations under international treaties, and property and debts. Russia has a multifaceted foreign policy. As of 2009, it maintains diplomatic relations with 191 countries and has 144 embassies. The foreign policy is determined by the President and implemented by the Ministry of Foreign Affairs of Russia.[79]

As the successor to a former superpower, Russia's geopolitical status has been often debated, particularly in relation to unipolar and multipolar views on the global political system. While Russia is commonly accepted to be a great power, in recent years it has been characterized by a number of world leaders,[80] [81] scholars,[82] commentators and politicians[83] as a currently reinstating or potential superpower.[84] [85] [86]

An important aspect of Russia's relations with the West is the criticism of Russia's political system and human rights management by the Western governments, the mass media and the leading democracy and human rights watchdogs.

In particular, such organisations as the Amnesty International and Human Rights Watch consider Russia to have not enough democratic attributes and to allow few political rights and civil liberties to its citizens.[87] [88] Freedom

House, an international organisation funded by the United States, ranks Russia as "not free", citing "carefully engineered elections" and "absence" of debate.[89] Russian authorities dismiss these claims and especially criticise Freedom House. The Russian Ministry of Foreign Affairs has called the 2006 *Freedom in the World* report "prefabricated", stating that the human rights issues have been turned into a political weapon in particular by the United States. The ministry also claims that such organisations as Freedom House and Human Rights Watch use the same scheme of voluntary extrapolation of *"isolated facts that of course can be found in any country"* into *"dominant tendencies"*.[90]

Member states, observers and partners of the Shanghai Cooperation Organisation

As one of five permanent members of the UN Security Council, Russia plays a major role in maintaining international peace and security. The country participates in the Quartet on the Middle East and the Six-party talks with North Korea. Russia is a member of the G8 industrialized nations, the Council of Europe, OSCE and APEC. Russia usually takes a leading role in regional organisations such as the CIS, EurAsEC, CSTO, and the SCO.[91] Former President Vladimir Putin had advocated a strategic partnership with close integration in various dimensions including establishment of EU-Russia Common Spaces.[92] Since the dissolution of the Soviet Union, Russia has developed a friendlier relationship with the United States and NATO. The NATO-Russia Council was established in 2002 to allow the United States, Russia and the 25 allies in NATO to work together as equal partners to pursue opportunities for joint collaboration.[93]

Russia maintains strong and positive relations with other BRIC countries. In recent years, the country has sought to strengthen ties especially with the People's Republic of China by signing the Treaty of Friendship as well as building the Trans-Siberian oil pipeline geared toward growing Chinese energy needs.[94]

Military

The Russian military is divided into the Ground Forces, Navy, and Air Force. There are also three independent arms of service: Strategic Missile Troops, Russian Space Forces, and the Airborne Troops. In 2006, the military had 1.037 million personnel on active duty.[] It is mandatory for all male citizens aged 18–27 to be drafted for a year of service in Armed Forces.[17]

2010 Moscow Victory Day Parade

Russia has the largest stockpile of nuclear weapons in the world. It has the second largest fleet of ballistic missile submarines and is the only country apart from the U.S. with a modern strategic bomber force.[] [95] Russia's tank force is the largest in the world, its surface navy and air force are among the largest ones.

The country has a large and fully indigenous arms industry, producing most of its own military equipment with only few types of weapons imported. Russia is the world's top supplier of arms, a spot it has held since 2001, accounting for around 30% of worldwide weapons sales[96] and exporting weapons to about 80 countries.[97]

Official government military spending for 2012 was $71.9 billion, the third largest in the world, though various sources have estimated Russia's military expenditures to be considerably higher.[] [98] Currently, a major equipment upgrade worth about $200 billion is on its way between 2006 and 2015.[99]

According to 2012 Global Peace Index, Russia is the sixth least peaceful country in the world, principally because of its defense industry. Russia has historically ranked low on the index since its inception in 2007.[100]

Political divisions

Federal subjects

The Russian Federation comprises 83 federal subjects.[101] These subjects have equal representation—two delegates each—in the Federation Council.[102] However, they differ in the degree of autonomy they enjoy.

- 46 oblasts (provinces): most common type of federal subjects, with locally elected[103] governor and legislature.
- 21 republics: nominally autonomous; each has its own constitution, direct-elected[103] head of republic[104] or a similar post, and parliament. Republics are allowed to establish their own official language alongside Russian but are represented by the federal government in international affairs. Republics are meant to be home to specific ethnic minorities.
- 9 krais (territories): essentially the same as oblasts. The "territory" designation is historic, originally given to frontier regions and later also to the administrative divisions that comprised autonomous okrugs or autonomous oblasts.
- 4 autonomous okrugs (autonomous districts): originally autonomous entities within oblasts and krais created for ethnic minorities, their status was elevated to that of federal subjects in the 1990s. With the exception of Chukotka Autonomous Okrug, all autonomous okrugs are still administratively subordinated to a krai or an oblast of which they are a part.
- 1 autonomous oblast (the Jewish Autonomous Oblast): historically, autonomous oblasts were administrative units subordinated to krais. In 1990, all of them except for the Jewish AO were elevated in status to that of a republic.
- 2 federal cities (Moscow and St. Petersburg): major cities that function as separate regions.

Federal districts

Federal subjects are grouped into eight federal districts, each administered by an envoy appointed by the President of Russia.[105] Unlike the federal subjects, the federal districts are not a subnational level of government, but are a level of administration of the federal government. Federal districts' envoys serve as liaisons between the federal subjects and the federal government and are primarily responsible for overseeing the compliance of the federal subjects with the federal laws.

Geography

Russia is the largest country in the world; its total area is 17075400 square kilometres (sq mi). There are 23 UNESCO World Heritage Sites in Russia, 40 UNESCO biosphere reserves,[106] 41 national parks and 101 nature reserves. It lies between latitudes 41° and 82° N, and longitudes 19° E and 169° W.

Russia has a wide natural resource base, including major deposits of timber, petroleum, natural gas, coal, ores and other mineral resources.

The topography of Russia

Topography

The two widest separated points in Russia are about 8000 km (mi) apart along a geodesic line. These points are: the boundary with Poland on a 60 km (mi) long Vistula Spit separating the Gdańsk Bay from the Vistula Lagoon; and the farthest southeast of the Kuril Islands. The points which are furthest separated in longitude are 6600 km (mi) apart along a geodesic line. These points are: in the west, the same spit; in the east, the Big Diomede Island. The Russian Federation spans 9 time zones.

Mount Elbrus, the highest point of the Caucasus, Russia and Europe

Most of Russia consists of vast stretches of plains that are predominantly steppe to the south and heavily forested to the north, with tundra along the northern coast. Russia possesses 10% of the world's arable land.[107] Mountain ranges are found along the southern borders, such as the Caucasus (containing Mount Elbrus, which at 5642 m (ft) is the highest point in both Russia and Europe) and the Altai (containing Mount Belukha, which at the 4506 m (ft) is the highest point of Siberia outside of the Russian Far East); and in the eastern parts, such as the Verkhoyansk Range or the volcanoes of Kamchatka Peninsula (containing Klyuchevskaya Sopka, which at the 4750 m (ft) is the highest active volcano in Eurasia as well as the highest point of Asian Russia). The Ural Mountains, rich in mineral resources, form a north-south range that divides Europe and Asia.

Russia has an extensive coastline of over 37000 km (mi) along the Arctic and Pacific Oceans, as well as along the Baltic Sea, Sea of Azov, Black Sea and Caspian Sea.[17] The Barents Sea, White Sea, Kara Sea, Laptev Sea, East Siberian Sea, Chukchi Sea, Bering Sea, Sea of Okhotsk, and the Sea of Japan are linked to Russia via the Arctic and Pacific. Russia's major islands and archipelagos include Novaya Zemlya, the Franz Josef Land, the Severnaya Zemlya, the New Siberian Islands, Wrangel Island, the Kuril Islands, and Sakhalin. The Diomede Islands (one controlled by Russia, the other by the U.S.) are just 3 km (mi) apart, and Kunashir Island is about 20 km (mi) from Hokkaido, Japan.

Russia has thousands of rivers and inland bodies of water, providing it with one of the world's largest surface water resources. Its lakes contain approximately one-quarter of the world's liquid fresh water.[] The largest and most prominent of Russia's bodies of fresh water is Lake Baikal, the world's deepest, purest, oldest and most capacious fresh water lake.[] Baikal alone contains over one-fifth of the world's fresh surface water.[] Other major lakes include Ladoga and Onega, two of the largest lakes in Europe. Russia is second only to Brazil in volume of the total renewable water resources. Of the country's 100,000 rivers,[108] the Volga is the most famous, not only because it is the longest river in Europe, but also because of its major role in Russian history.[17] The Siberian rivers Ob, Yenisey, Lena and Amur are among the longest rivers in the world.

Central Russian Upland near Zaraysk, Moscow Oblast

Climate

The enormous size of Russia and the remoteness of many areas from the sea result in the dominance of the humid continental climate, which is prevalent in all parts of the country except for the tundra and the extreme southeast. Mountains in the south obstruct the flow of warm air masses from the Indian Ocean, while the plain of the west and north makes the country open to Arctic and Atlantic influences.[]

Taiga forest in winter, Arkhangelsk Oblast

Most of Northern European Russia and Siberia has a subarctic climate, with extremely severe winters in the inner regions of Northeast Siberia (mostly the Sakha Republic, where the Northern Pole of Cold is located with the record low temperature of), and more moderate elsewhere. The strip of land along the shore of the Arctic Ocean, as well as the Russian Arctic islands, have a polar climate.

The coastal part of Krasnodar Krai on the Black Sea, most notably in Sochi, possesses a humid subtropical climate with mild and wet winters. Winter is dry compared to summer in many regions of East Siberia and the Far East, while other parts of the country experience more even precipitation across seasons. Winter precipitation in most parts of the country usually falls as snow. The region along the Lower Volga and Caspian Sea coast, as well as some areas of southernmost Siberia, possesses a semi-arid climate.

Throughout much of the territory there are only two distinct seasons—winter and summer; spring and autumn are usually brief periods of change between extremely low temperatures and extremely high.[] The coldest month is January (February on the coastline), the warmest usually is July. Great ranges of temperature are typical. In winter, temperatures get colder both from south to north and from west to east. Summers can be quite hot, even in Siberia.[109] The continental interiors are the driest areas.

Biodiversity

From north to south the East European Plain, also known as Russian Plain, is clad sequentially in Arctic tundra, coniferous forest (taiga), mixed and broad-leaf forests, grassland (steppe), and semi-desert (fringing the Caspian Sea), as the changes in vegetation reflect the changes in climate. Siberia supports a similar sequence but is largely taiga. Russia has the world's largest forest reserves,[] known as "the lungs of Europe",[110] second only to the Amazon Rainforest in the amount of carbon dioxide it absorbs.

There are 266 mammal species and 780 bird species in Russia. A total of 415 animal species have been included in the Red Data Book of the Russian Federation as of 1997[111] and are now protected.

The brown bear is a popular symbol of Russia, particularly in the West.

Economy

World Trade Center in Moscow

Russia has a market economy with enormous natural resources, particularly oil and natural gas. It has the 8th largest economy in the world by nominal GDP and the 6th largest by purchasing power parity (PPP). Since the turn of the 21st century, higher domestic consumption and greater political stability have bolstered economic growth in Russia. The country ended 2008 with its ninth straight year of growth, averaging 7% annually between 2000 and 2008. Real GDP per capita, PPP (current international $) was 19,840 in 2010.[] Growth was primarily driven by non-traded services and goods for the domestic market, as opposed to oil or mineral extraction and exports.[17] The average nominal salary in Russia was $640 per month in early 2008, up from $80 in 2000.[112] In the end of 2011 the average nominal monthly wages reached 24,310 RUR (or US$810),[113] while tax on the income of individuals is payable at the rate of 13% on most incomes.[114] Approximately 13.7% of Russians lived below the national poverty line in 2010,[115] significantly down from 40% in 1998 at the worst point of the post-Soviet collapse.[] Unemployment in Russia was at 6% in 2007, down from about 12.4% in 1999.[116] The middle class has grown from just 8 million persons in 2000 to 55 million persons in 2006.[117] Sugar imports reportedly dropped 82% between 2012–2013.[118]

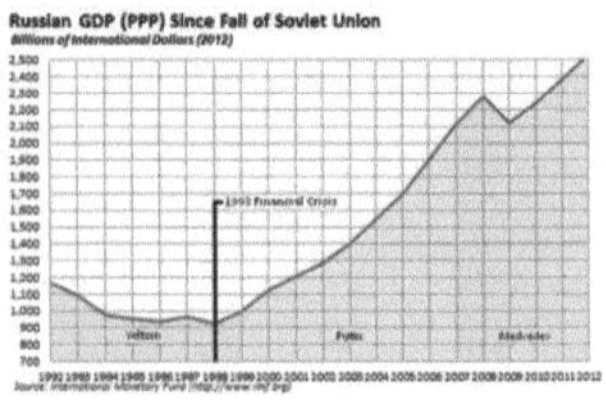

Russian economy since the end of the Soviet Union

Oil, natural gas, metals, and timber account for more than 80% of Russian exports abroad.[17] Since 2003, the exports of natural resources started decreasing in economic importance as the internal market strengthened considerably. Despite higher energy prices, oil and gas only contribute to 5.7% of Russia's GDP and the government predicts this will be 3.7% by 2011.[119] Oil export earnings allowed Russia to increase its foreign reserves from $12 billion in 1999 to $597.3 billion on 1 August 2008, the third largest foreign exchange reserves in the world.[120] The macroeconomic policy under Finance Minister Alexei Kudrin was prudent and sound, with excess income being stored in the

Stabilization Fund of Russia.[] In 2006, Russia repaid most of its formerly massive debts,[121] leaving it with one of the lowest foreign debts among major economies.[122] The Stabilization Fund helped Russia to come out of the global financial crisis in a much better state than many experts had expected.[]

A simpler, more streamlined tax code adopted in 2001 reduced the tax burden on people and dramatically increased state revenue.[123] Russia has a flat tax rate of 13%. This ranks it as the country with the second most attractive personal tax system for single managers in the world after the United Arab Emirates.[124] According to Bloomberg, Russia is considered well ahead of most other resource-rich countries in its economic development, with a long tradition of education, science, and industry.[125] The country has a higher proportion of higher education graduates than any other country in Eurasia.[126]

The economic development of the country has been uneven geographically with the Moscow region contributing a very large share of the country's GDP.[127] Another problem is modernisation of infrastructure, ageing and inadequate after years of being neglected in 1990s; the government has said $1 trillion will be invested in development of infrastructure by 2020.[128] In December 2011, Russia finally joined World Trade Organisation, allowing it a greater access to overseas markets. Some analysts estimate that WTO membership could bring the Russian economy a bounce of up to 3 per cent annually.[129] Russia ranks the second most corrupt country in Europe (after Ukraine), according to the Corruption Perceptions Index. The Norwegian-Russian Chamber of Commerce also states that "*[c]orruption is one of the biggest problems both Russian and international companies have to deal with.*"[130]

Agriculture

The total area of cultivated land in Russia was estimated as 1,237,294 km^2 in 2005, the fourth largest in the world.[131] In 1999–2009, Russia's agriculture demonstrated steady growth,[132] and the country turned from a grain importer to the third largest grain exporter after EU and USA.[133] The production of meat has grown from 6,813,000 tonnes in 1999 to 9,331,000 tonnes in 2008, and continues to grow.[134]

Rye Fields, by Ivan Shishkin. Russia is the world's top producer of rye, barley, buckwheat, oats and sunflower seed, and one of the largest producers and exporters of wheat.

This restoration of agriculture was supported by credit policy of the government, helping both individual farmers and large privatized corporate farms, that once were Soviet kolkhozes and still own the significant share of agricultural land.[135] While large farms concentrate mainly on the production of grain and husbandry products, small private household plots produce most of the country's yield of potatoes, vegetables and fruits.[136]

With access to three of the world's oceans—the Atlantic, Arctic, and Pacific—Russian fishing fleets are a major contributor to the world's fish supply. The total capture of fish was at 3,191,068 tons in 2005.[137] Both exports and imports of fish and sea products grew significantly in the recent years, reaching correspondingly $2,415 and $2,036 millions in 2008.[138]

Sprawling from the Baltic Sea to the Pacific Ocean, Russia has more than a fifth of the world's forests, which makes it the largest forest country in the world.[] [139] However, according to a 2012 study by the Food and Agriculture Organization of the United Nations and the Government of the Russian Federation,[140] the considerable potential of Russian forests is underutilized and Russia's share of the global trade in forest products is less than four percent.[141] [142]

Energy

Russia is a key oil and gas supplier to much of Europe.

In recent years, Russia has frequently been described in the media as an energy superpower.[143] [144] The country has the world's largest natural gas reserves,[145] the 8th largest oil reserves,[146] and the second largest coal reserves.[147] Russia is the world's leading natural gas exporter[148] and second largest natural gas producer,[17] while also the largest oil exporter and the largest oil producer.[16] On 1 January 2011, Russia said it had begun scheduled oil shipments to China, with the plan to increase the rate up to 300,000 barrels per day in 2011.[149]

Russia is the 3rd largest electricity producer in the world[150] and the 5th largest renewable energy producer, the latter because of the well-developed hydroelectricity production in the country.[151] Large cascades of hydropower plants are built in European Russia along big rivers like Volga. The Asian part of Russia also features a number of major hydropower stations, however the gigantic hydroelectric potential of Siberia and the Russian Far East largely remains unexploited.

Russia was the first country to develop civilian nuclear power and to construct the world's first nuclear power plant. Currently the country is the 4th largest nuclear energy producer,[152] with all nuclear power in Russia being managed by Rosatom State Corporation. The sector is rapidly developing, with an aim of increasing the total share of nuclear energy from current 16.9% to 23% by 2020. The Russian government plans to allocate 127 billion rubles ($5.42 billion) to a federal program dedicated to the next generation of nuclear energy technology. About 1 trillion rubles ($42.7 billion) is to be allocated from the federal budget to nuclear power and industry development before 2015.[153]

Transport

Railway transport in Russia is mostly under the control of the state-run Russian Railways monopoly. The company accounts for over 3.6% of Russia's GDP and handles 39% of the total freight traffic (including pipelines) and more than 42% of passenger traffic.[] The total length of common-used railway tracks exceeds 85,500 km,[] second only to the U.S. Over 44,000 km of tracks are electrified,[154] which is the largest number in the world, and additionally there are more than 30,000 km of industrial non-common carrier lines. Railways in Russia, unlike in the most of the world, use broad gauge of , with the exception of 957 km on Sakhalin island using narrow gauge of . The most renown railway in Russia is Trans-Siberian (*Transsib*), spanning a record 7 time zones and serving the longest single continuous services in the world, Moscow-Vladivostok (9,259 km, 5,753 mi), Moscow–Pyongyang (10,267 km, 6,380 mi)[155] and Kiev–Vladivostok (11,085 km, 6,888 mi).[156]

The marker for kilometre 9288, at the end of the Trans-Siberian Railway in Vladivostok

As of 2006 Russia had 933,000 km of roads, of which 755,000 were paved.[157] Some of these make up the Russian federal motorway system. With a large land area the road density is the lowest of all the G8 and BRIC countries.[158]

Much of Russia's inland waterways, which total 102000 km (mi), are made up of natural rivers or lakes. In the European part of the country the network of channels connects the basins of major rivers. Russia's capital, Moscow, is sometimes called *"the port of the five seas"*, because of its waterway connections to the Baltic, White, Caspian, Azov and Black Seas.

Major sea ports of Russia include Rostov-on-Don on the Azov Sea, Novorossiysk on the Black Sea, Astrakhan and Makhachkala on the Caspian, Kaliningrad and St Petersburg on the Baltic, Arkhangelsk on the White Sea, Murmansk on the Barents Sea, Petropavlovsk-Kamchatsky and Vladivostok on the Pacific Ocean. In 2008 the country owned 1448 merchant marine ships. The world's only fleet of nuclear-powered icebreakers advances the economic exploitation of the Arctic continental shelf of Russia and the development of sea trade through the Northern Sea Route between Europe and East Asia.

Yamal, one of Russia's nuclear-powered icebreakers (gallery [159])

By total length of pipelines Russia is second only to the U.S. Currently many new pipeline projects are being realized, including Nord Stream and South Stream natural gas pipelines to Europe, and the Eastern Siberia – Pacific Ocean oil pipeline (ESPO) to the Russian Far East and China.

Russia has 1216 airports,[160] the busiest being Sheremetyevo, Domodedovo, and Vnukovo in Moscow, and Pulkovo in St Petersburg. The total length of runways in Russia exceeds **unknown operator: u','** kilometres (**unknown operator: u'strong'** mi).[161]

Typically, major Russian cities have well-developed systems of public transport, with the most common varieties of exploited vehicles being bus, trolleybus and tram. Seven Russian cities, namely Moscow, Saint Petersburg, Nizhny Novgorod, Novosibirsk, Samara, Yekaterinburg and Kazan, have underground metros, while Volgograd features a metrotram. The total length of metros in Russia is 465.4 kilometres (mi). Moscow Metro and Saint Petersburg Metro are the oldest in Russia, opened in 1935 and 1955 respectively. These two are among the fastest and busiest metro systems in the world, and are famous for rich decorations and unique designs of their stations, which is a common tradition on Russian metros and railways.

Science and technology

Science and technology in Russia blossomed since the Age of Enlightenment, when Peter the Great founded the Russian Academy of Sciences and Saint Petersburg State University, and polymath Mikhail Lomonosov established the Moscow State University, paving the way for a strong native tradition in learning and innovation. In the 19th and 20th centuries the country produced a large number of notable scientists and inventors.

Mikhail Lomonosov, polymath scientist, inventor, poet and artist

The Russian physics school began with Lomonosov who proposed the law of conservation of matter preceding the energy conservation law. Russian discoveries and inventions in physics include the electric arc, electrodynamical Lenz's law, space groups of crystals, photoelectric cell, Cherenkov radiation, electron paramagnetic resonance, heterotransistors and 3D holography. Lasers and masers were co-invented by Nikolai Basov and Alexander Prokhorov, while the idea of tokamak for controlled nuclear fusion was introduced by Igor Tamm, Andrei Sakharov and Lev Artsimovich, leading eventually the modern international ITER project, where Russia is a party.

Since the time of Nikolay Lobachevsky (a *Copernicus of Geometry* who pioneered the non-Euclidean geometry) and a prominent tutor Pafnuty Chebyshev, the Russian mathematical school became one of the most influential in the world.[162] Chebyshev's students included Aleksandr Lyapunov, who founded the modern stability theory, and

Andrey Markov who invented the Markov chains. In the 20th century Soviet mathematicians, such as Andrey Kolmogorov, Israel Gelfand and Sergey Sobolev, made major contributions to various areas of mathematics. Nine Soviet/Russian mathematicians were awarded with Fields Medal, a most prestigious award in mathematics. Recently Grigori Perelman was offered the first ever Clay Millennium Prize Problems Award for his final proof of the Poincaré conjecture in 2002.[163]

Russian chemist Dmitry Mendeleev invented the Periodic table, the main framework of modern chemistry. Aleksandr Butlerov was one of the creators of the theory of chemical structure, playing a central role in organic chemistry. Russian biologists include Dmitry Ivanovsky who discovered viruses, Ivan Pavlov who was the first to experiment with the classical conditioning, and Ilya Mechnikov who was a pioneer researcher of the immune system and probiotics.

Soyuz TMA-2 is launched from Baikonur, Kazakhstan, carrying one of the first resident crews to the International Space Station

Many Russian scientists and inventors were émigrés, like Igor Sikorsky, who built the first airliners and modern-type helicopters; Vladimir Zworykin, often called the father of TV; chemist Ilya Prigogine, noted for his work on dissipative structures and complex systems; Nobel Prize-winning economists Simon Kuznets and Wassily Leontief; physicist Georgiy Gamov (an author of the Big Bang theory) and social scientist Pitirim Sorokin. Many foreigners worked in Russia for a long time, like Leonard Euler and Alfred Nobel.

Russian inventions include the arc welding by Nikolay Benardos, further developed by Nikolay Slavyanov, Konstantin Khrenov and other Russian engineers. Gleb Kotelnikov invented the knapsack parachute, while Evgeniy Chertovsky introduced the pressure suit. Alexander Lodygin and Pavel Yablochkov were pioneers of electric lighting, and Mikhail Dolivo-Dobrovolsky introduced the first three-phase electric power systems, widely used today. Sergei Lebedev invented the first commercially viable and mass-produced type of synthetic rubber. The first ternary computer, *Setun*, was developed by Nikolay Brusentsov.

In the 20th century a number of prominent Soviet aerospace engineers, inspired by the fundamental works of Nikolai Zhukovsky, Sergei Chaplygin and others, designed many hundreds of models of military and civilian aircraft and founded a number of *KBs* (*Construction Bureaus*) that now constitute the bulk of Russian United Aircraft Corporation. Famous Russian aircraft include the civilian Tu-series, Su and MiG fighter aircraft, Ka and Mi-series helicopters; many Russian aircraft models are on the list of most produced aircraft in history.

Famous Russian battle tanks include T-34, the best tank design of World War II,[164] and further tanks of T-series, including the most produced tank in history, T-54/55.[165] The AK-47 and AK-74 by Mikhail Kalashnikov constitute the most widely used type of assault rifle throughout the world—so much so that more AK-type rifles have been manufactured than all other assault rifles combined.[166]

The Sukhoi PAK FA is a fifth-generation jet fighter being developed for the Russian Air Force.

Yo-mobil, a new Russian hybrid car

With all these achievements, however, since the late Soviet era Russia was lagging behind the West in a number of technologies, mostly those related to energy conservation and consumer goods production. The crisis of 1990-s led to the drastic reduction of the state support for science and a brain drain migration from Russia.

In the 2000s, on the wave of a new economic boom, the situation in the Russian science and technology has improved, and the government launched a campaign aimed into modernisation and innovation. Russian President Dmitry Medvedev formulated top 5 priorities for the country's technological development: efficient energy use, IT (including both common products and the products combined with space technology), nuclear energy and pharmaceuticals.[167]

Currently Russia has completed the GLONASS satellite navigation system. The country is developing its own fifth-generation jet fighter and constructing the first serial mobile nuclear plant in the world. In 2010, an economy class hybrid electric car project was introduced, called Yo-mobil, that will be mass-produced by ё-Auto, a Russian company that is a joint venture between truck maker Yarovit and the Onexim investment group.[] [] []

Space exploration

Soviet and Russian space station Mir

Russian achievements in the field of space technology and space exploration are traced back to Konstantin Tsiolkovsky, the father of theoretical austronautics.[168] His works had inspired leading Soviet rocket engineers, such as Sergey Korolyov, Valentin Glushko and many others who contributed to the success of the Soviet space program on early stages of the Space Race and beyond.

In 1957 the first Earth-orbiting artificial satellite, *Sputnik 1*, was launched; in 1961 the first human trip into space was successfully made by Yury Gagarin. Many other Soviet and Russian space exploration records ensued, including the first spacewalk performed by Alexey Leonov, Luna 9 was the first spacecraft to land on the Moon, Venera 7 was the first to land on another planet (Venus), Mars 3 then the first to land on Mars, the first space exploration rover *Lunokhod-1* and the first space station *Salyut 1* and *Mir*.

After the collapse of the Soviet Union, some government-funded space exploration programs, including the Buran space shuttle program, were cancelled or delayed, while participation of the Russian space industry in commercial activities and international cooperation intensified. Nowadays Russia is the largest satellite launcher.[169] After the U.S. Space Shuttle program ended in 2011, Soyuz rockets became the only provider of transport for astronauts at the International Space Station.

Demographics

Percentage of ethnic Russians by region in 2010

>80%

70—79%

50—69%

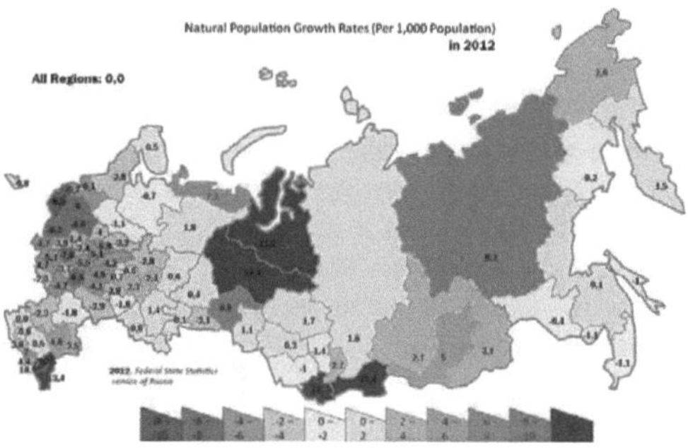

Natural population growth rate in Russia, 2012.

Ethnic Russians comprise 81% of the country's population.[170] The Russian Federation is also home to several sizeable minorities. In all, 160 different other ethnic groups and indigenous peoples live within its borders.[171] Though Russia's population is comparatively large, its density is low because of the country's enormous size. Population is densest in European Russia, near the Ural Mountains, and in southwest Siberia. 73% of the population lives in urban areas while 27% in rural ones.[172] The results of the 2010 Census show a total population of 142,856,536.[]

Russia's population peaked at 148,689,000 in 1991, just before the dissolution of the Soviet Union. It began to experience a rapid decline starting in the mid-1990s.[173] The decline has slowed to near stagnation in recent years because of reduced death rates, increased birth rates and increased immigration.[174]

In 2009, Russia recorded annual population growth for the first time in fifteen years, with total growth of 10,500.[174] 279,906 migrants arrived to the Russian Federation the same year, of which 93% came from CIS countries.[174] The number of Russian emigrants steadily declined from 359,000 in 2000 to 32,000 in 2009.[174] There are also an estimated 10 million illegal immigrants from the ex-Soviet states in Russia.[175] Roughly 116 million ethnic Russians live in Russia[171] and about 20 million more live in other former republics of the Soviet Union,[176] mostly in Ukraine and Kazakhstan.[177]

The 2010 census recorded 81% of the population as ethnically Russian, and 19% as other ethnicities:[170] 3.7% Tatars; 1.4% Ukrainians; 1.1% Bashkirs; 1% Chuvashes; 11.8% others and unspecified. According to the Census,

84.93% of the Russian population belongs to European ethnic groups (Slavic, Germanic, Finnic other than Ugric, Greek.etc.). This is a decline from the 2002, when they constituted for more than 86% of the population.[170]

Russia's birth rate is higher than that of most European countries (12.6 births per 1000 people in 2010[174] compared to the European Union average of 9.90 per 1000),[178] but its death rate is also substantially higher (in 2010, Russia's death rate was 14.3 per 1000 people[174] compared to the EU average of 10.28 per 1000).[179] The Russian Ministry of Health and Social Affairs predicted that by 2011 the death rate would equal the birth rate because of increase in fertility and decline in mortality.[180] The government is implementing a number of programs designed to increase the birth rate and attract more migrants. Monthly government child assistance payments were doubled to US$55, and a one-time payment of US$9,200 was offered to women who had a second child since 2007.[181] In 2009 Russia experienced its highest birth rate since the dissolution of the Soviet Union.[174] [182]

Language

Russia's 160 ethnic groups speak some 100 languages.[] According to the 2002 Census, 142.6 million people speak Russian, followed by Tatar with 5.3 million and Ukrainian with 1.8 million speakers.[183] Russian is the only official state language, but the Constitution gives the individual republics the right to make their native language official in addition to Russian.[184]

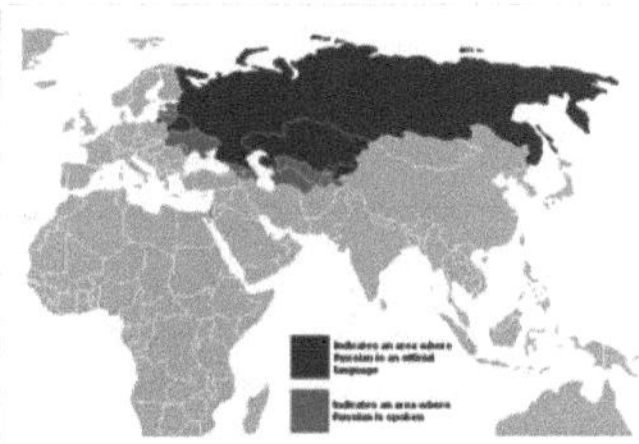
Countries where the Russian language is spoken

Despite its wide distribution, the Russian language is homogeneous throughout the country. Russian is the most geographically widespread language of Eurasia, as well as the most widely spoken Slavic language.[185] It belongs to the Indo-European language family and is one of the living members of the East Slavic languages, the others being Belarusian and Ukrainian (and possibly Rusyn). Written examples of Old East Slavic (*Old Russian*) are attested from the 10th century onwards.[186]

Russian is one of the six official languages of the UN.[187]

Religion

The Temple of All Religions in Kazan

Orthodox Christianity, Islam, Judaism and Buddhism are Russia's traditional religions, and are all legally a part of Russia's "historical heritage".[188] Traced back to the Christianization of Kievan Rus' in the 10th century, Russian Orthodoxy is the dominant religion in the country; smaller Christian denominations such as Catholics, Armenian Gregorians and various Protestant churches also exist.. The Russian Orthodox Church was the country's state religion prior to the Revolution and remains the largest religious body in the country. Estimates on how many believers the church has fluctuate widely in different sources. A 15 January 2013 statistic by the Russian Academy of Sciences announced that 79% of the Russian population belong to the Russian Orthodox Church.[] 95% of the registered Orthodox parishes belong to the Russian Orthodox Church while there are a number of smaller Orthodox Churches.[189] However, the vast majority of Orthodox believers do not attend church on a regular basis. Easter is the most popular religious holiday in Russia, celebrated by more than 90% of all Russian citizens, including large numbers of those who are non-religious. More than three-quarters of the Russian population celebrate Easter by making traditional Easter cakes, coloured eggs and paskha.[190]

Estimates of the number of Muslims in Russia range from 7–9 million by local sources, to 15–20 million by Western and Islamic sources.[191] There are approximately 3 to 4 million temporary Muslim migrants from the post-Soviet states.[192] Most Muslims live in the Volga-Ural region, as well as in the Caucasus, Moscow, St. Petersburg and Western Siberia.[193]

Buddhism is traditional in three regions of the Russian Federation: Buryatia, Tuva, and Kalmykia. Some residents of the Siberian and Far Eastern regions, such as Yakutia and Chukotka, practice shamanist, pantheistic, and pagan rites, along with the major religions. Induction into religion takes place primarily along ethnic lines. Slavs are overwhelmingly Orthodox Christian, Turkic speakers are predominantly Muslim, and Mongolic peoples are generally Buddhists.[194]

Various reports put the number of non-religious in Russia at between 16–48% of the population.[195] The number of atheists has decreased significantly; according to the recent statistic, only seven percent declared themselves atheists, a decrease of 5% in three years.[]

Health

The Russian Constitution guarantees free, universal health care for all its citizens.[196] In practice, however, free health care is partially restricted because of mandatory registration.[197] While Russia has more physicians, hospitals, and health care workers than almost any other country in the world on a per capita basis,[198] since the dissolution of the Soviet Union the health of the Russian population has declined considerably as a result of social, economic, and lifestyle changes;[199] the trend has been reversed only in the recent years, with average life expectancy having increased 2.4 years for males and 1.4 years for females between 2006–09.[174]

A mobile clinic used to provide health care at remote railway stations

As of 2009, the average life expectancy in Russia was 62.77 years for males and 74.67 years for females.[200] The biggest factor contributing to the relatively low life expectancy for males is a high mortality rate among working-age males. Deaths mostly occur because of preventable causes (e.g., alcohol poisoning, smoking, traffic accidents, violent crime).[174] As a result of the large gender difference in life expectancy, and also because of the lasting effect of high casualties in World War II, the gender imbalance remains to this day and there are 0.859 males to every female.[17]

Education

A school in Moscow. The tower of the Moscow State University is seen at the distance.

Russia has a free education system – this is guaranteed for all citizens by the Constitution,[201] however an entry to subsidized post-secondary education is highly competitive.[202] As a result of great emphasis on science and technology in education, Russian medical, mathematical, scientific, and aerospace research is generally of a high order.[203]

Since 1990, the 11-year school education has been introduced. Education in state-owned secondary schools is free. University level education is free, with exceptions. A substantial share of students is enrolled for full pay (many state institutions started to open commercial positions in the last years).[204]

In 2004, state spending for education amounted to 3.6% of the GDP, or 13% of the consolidated state budget.[205] The Government allocates funding to pay the tuition fees within an established quota or number of students for each state institution. In higher education institutions, students are paid a small stipend and provided with free housing if they are from out of town.[206]

The oldest and largest Russian universities are Moscow State University and Saint Petersburg State University. In the 2000s, in order to create higher education and research institutions of comparable scale in Russian regions, the government launched a program of establishing *federal universities*, mostly by merging existing large regional universities and research institutes and providing them with a special funding. These new institutions include the Southern Federal University, Siberian Federal University, Kazan Volga Federal University, North-Eastern Federal University and Far Eastern Federal University.

Culture

Folk culture and cuisine

There are over 160 different ethnic groups and indigenous peoples in Russia.[171] Ethnic Russians with their Slavic Orthodox traditions, Tatars and Bashkirs with their Turkic Muslim culture, Buddhist nomadic Buryats and Kalmyks, Shamanistic peoples of the Extreme North and Siberia, highlanders of the Northern Caucasus, Finno-Ugric peoples of the Russian North West and Volga Region all contribute to the cultural diversity of the country.

The Merchant's Wife by Boris Kustodiev, showcasing the Russian tea culture

Handicraft, like Dymkovo toy, khokhloma, gzhel and palekh miniature represent an important aspect of Russian folk culture. Ethnic Russian clothes include kaftan, kosovorotka and ushanka for men, sarafan and kokoshnik for women, with lapti and valenki as common shoes. The clothes of Cossacks from Southern Russia include burka and papaha, which they share with the peoples of the Northern Caucasus.

Russian cuisine widely uses fish, poultry, mushrooms, berries, and honey. Crops of rye, wheat, barley, and millet provide the ingredients for various breads, pancakes and cereals, as well as for kvass, beer and vodka drinks. Black bread is rather popular in Russia, compared to the rest of the world. Flavourful soups and stews include shchi, borsch, ukha, solyanka and okroshka. Smetana (a heavy sour cream) is often added to soups and salads. Pirozhki, blini and syrniki are native types of pancakes. Chicken Kiev, pelmeni and shashlyk are popular meat dishes, the last two being of Tatar and Caucasus origin respectively. Other meat dishes include stuffed cabbage rolls *(golubtsy)* usually filled with meat.[] Salads include Olivier salad, Vinaigrette and Dressed herring.

Russia's large number of ethnic groups have distinctive traditions regarding folk music. Typical ethnic Russian musical instruments are gusli, balalaika, zhaleika and garmoshka. Folk music had a significant influence on Russian classical composers, and in modern times it is a source of inspiration for a number of popular folk bands, like Melnitsa. Russian folk songs, as well as patriotic Soviet songs, constitute the bulk of the repertoire of the world-renown Red Army choir and other popular ensembles.

Russians have many traditions, including the washing in banya, a hot steam bath somewhat similar to sauna.[38] Old Russian folklore takes its roots in the pagan Slavic religion. Many Russian fairy tales and epic bylinas were adaptated for animation films, or for feature movies by the prominent directors like Aleksandr Ptushko (*Ilya Muromets*, *Sadko*) and Aleksandr Rou (*Morozko*, *Vasilisa the Beautiful*). Russian poets, including Pyotr Yershov and Leonid Filatov, made a number of well-known poetical interpretations of the classical fairy tales, and in some cases, like that of Alexander Pushkin, also created fully original fairy tale poems of great popularity.

Architecture

Saint Basil's Cathedral in Moscow, a well known piece of Russian architecture

Since the Christianization of Kievan Rus' for several ages Russian architecture was influenced predominantly by the Byzantine architecture. Apart from fortifications (kremlins), the main stone buildings of ancient Rus' were Orthodox churches with their many domes, often gilded or brightly painted.

Aristotle Fioravanti and other Italian architects brought Renaissance trends into Russia since the late 15th century, while the 16th century saw the development of unique tent-like churches[207] culminating in Saint Basil's Cathedral. By that time the onion dome design was also fully developed.[208] In the 17th century, the "fiery style" of ornamentation flourished in Moscow and Yaroslavl, gradually paving the way for the Naryshkin baroque of the 1690s. After the reforms of Peter the Great the change of architectural styles in Russia generally followed that in the Western Europe.

The 18th-century taste for rococo architecture led to the ornate works of Bartolomeo Rastrelli and his followers. The reigns of Catherine the Great and her grandson Alexander I saw the flourishing of Neoclassical architecture, most notably in the capital city of Saint Petersburg. The second half of the 19th century was dominated by the Neo-Byzantine and Russian Revival styles. Prevalent styles of the 20th century were the Art Nouveau, Constructivism, and the Stalin Empire style.

In 1955, a new Soviet leader, Nikita Khrushchev, condemned the "excesses" of the former academic architecture,[209] and the late Soviet era was dominated by plain functionalism in architecture. This helped somewhat to resolve the housing problem, but created a large quantity of buildings of low architectural quality, much in contrast with the previous bright styles. The situation improved in the recent two decades. Many temples demolished in Soviet times were rebuilt, and this process continues along with the restoration of various historical buildings destroyed in World War II. A total of 23,000 Orthodox churches have been rebuilt between 1991–2010, which effectively quadrapled the number of operating churches in Russia.[210] For many, mostly elderly and conservative inhabitants, the lost Russia is identified with the Soviets. For a more liberal and largely middle-aged part of the Russian population it is the pre-revolutionary empire.[]

Visual arts

Early Russian painting is represented in icons and vibrant frescos, the two genres inherited from Byzantium. As Moscow rose to power, Theophanes the Greek, Dionisius and Andrei Rublev became vital names associated with a distinctly Russian art.

The *Trinity* icon by Andrei Rublev

The Russian Academy of Arts was created in 1757[211] and gave Russian artists an international role and status. Ivan Argunov, Dmitry Levitzky, Vladimir Borovikovsky and other 18th century academicians mostly focused on portrait painting. In the early 19th century, when neoclassicism and romantism flourished, mythological and Biblical themes inspired many prominent painings, notably by Karl Briullov and Alexander Ivanov.

In the mid-19th century the *Peredvizhniki* (*Wanderers*) group of artists broke with the Academy and initiated a school of art liberated from academic restrictions.[212] These were mostly realist painters who captured Russian identity in landscapes of wide rivers, forests, and birch clearings, as well as vigorous genre scenes and robust portraits of their contemporaries. Some artists focused on depicting dramatic moments in Russian history, while others turned to social criticism, showing the conditions of the poor and caricaturing authority; critical realism flourished under the reign of Alexander II. Leading realists include Ivan Shishkin, Arkhip Kuindzhi, Ivan Kramskoi, Vasily Polenov, Isaac Levitan, Vasily Surikov, Viktor Vasnetsov, Ilya Repin and Boris Kustodiev.

The turn of the 20th century saw the rise of symbolist painting, represented by Mikhail Vrubel, Kuzma Petrov-Vodkin and Nicholas Roerich.

The Russian avant-garde was a large, influential wave of modernist art that flourished in Russia from approximately 1890 to 1930. The term covers many separate, but inextricably related art movements that occurred at the time, namely neo-primitivism, suprematism, constructivism, rayonism, and Russian Futurism. Notable artists from this era include El Lissitzky, Kazimir Malevich, Wassily Kandinsky, and Marc Chagall. Since 1930s the revolutionary ideas of the avant-garde clashed with the newly emerged conservative direction of socialist realism.

Soviet art produced works that were furiously patriotic and anti-fascist during and after the Great Patriotic War. Multiple war memorials, marked by a great restrained solemnity, were built throughout the country. Soviet artists often combined innovation with socialist realism, notably the sculptors Vera Mukhina, Yevgeny Vuchetich and Ernst Neizvestny.

Music and dance

The *Snowdance* scene from *The Nutcracker* ballet, composed by Pyotr Ilyich Tchaikovsky

Music in 19th century Russia was defined by the tension between classical composer Mikhail Glinka along with other members of The Mighty Handful, who embraced Russian national identity and added religious and folk elements to their compositions, and the Russian Musical Society led by composers Anton and Nikolay Rubinsteins, which was musically conservative. The later tradition of Pyotr Ilyich Tchaikovsky, one of the greatest composers of the Romantic era, was continued into the 20th century by Sergei Rachmaninoff.[213] World-renowned composers of the 20th century include Alexander Scriabin, Igor Stravinsky, Sergei Prokofiev, Dmitri Shostakovich and Alfred Schnittke.

Russian conservatories have turned out generations of famous soloists. Among the best known are violinists David Oistrakh and Gidon Kremer; cellist Mstislav Rostropovich; pianists Vladimir Horowitz, Sviatoslav Richter, and Emil Gilels; and vocalists Fyodor Shalyapin, Galina Vishnevskaya, Anna Netrebko and Dmitry Hvorostovsky.[214]

During the early 20th century, Russian ballet dancers Anna Pavlova and Vaslav Nijinsky rose to fame, and impresario Sergei Diaghilev and his Ballets Russes' travels abroad profoundly influenced the development of dance worldwide.[215] Soviet ballet preserved the perfected 19th century traditions,[216] and the Soviet Union's choreography schools produced many internationally famous stars, including Maya Plisetskaya, Rudolf Nureyev, and Mikhail Baryshnikov. The Bolshoi Ballet in Moscow and the Mariinsky Ballet in St Petersburg remain famous throughout the world.[217]

Modern Russian rock music takes its roots both in the Western rock and roll and heavy metal, and in traditions of the Russian bards of the Soviet era, such as Vladimir Vysotsky and Bulat Okudzhava.[218] Popular Russian rock groups include Mashina Vremeni, DDT, Aquarium, Alisa, Kino, Kipelov, Nautilus Pompilius, Aria, Grazhdanskaya Oborona, Splean and Korol i Shut. Russian pop music developed from what was known in the Soviet times as *estrada* into full-fledged industry, with some performers gaining wide international recognition, such as t.A.T.u., Nu Virgos and Vitas.

Literature and philosophy

Leo Tolstoy, novelist and philosopher

In the 18th century, during the era of Russian Enlightenment, the development of Russian literature was boosted by the works of Mikhail Lomonosov and Denis Fonvizin. By the early 19th century a modern native tradition had emerged, producing some of the greatest writers in Russian history. This period, known also as the Golden Age of Russian Poetry, began with Alexander Pushkin, who is considered the founder of the modern Russian literary language and often described as the "Russian Shakespeare".[219] It continued into the 19th century with the poetry of Mikhail Lermontov and Nikolay Nekrasov, dramas of Alexander Ostrovsky and Anton Chekhov, and the prose of Nikolai Gogol and Ivan Turgenev. Leo Tolstoy and Fyodor Dostoyevsky have been described by literary critics as the greatest novelists of all time.[220] [221]

By the 1880s, the age of the great novelists was over, and short fiction and poetry became the dominant genres. The next several decades became known as the Silver Age of Russian Poetry, when the previously dominant literary realism was replaced by symbolism. Leading authors of this era include such poets as Valery Bryusov, Vyacheslav Ivanov, Alexander Blok, Nikolay Gumilev and Anna Akhmatova, and novelists Leonid Andreyev, Ivan Bunin, and Maxim Gorky.

Russian philosophy blossomed in the 19th century, when it was defined initially by the opposition of Westernizers, advocating Western political and economical models, and Slavophiles, insisting on developing Russia as a unique civilization. The latter group includes Nikolai Danilevsky and Konstantin Leontiev, the founders of eurasianism. In its further development Russian philosophy was always marked by a deep connection to literature and interest in creativity, society, politics and nationalism; Russian cosmism and religious philosophy were other major areas. Notable philosophers of the late 19th and the early 20th centuries include Vladimir Solovyev, Sergei Bulgakov, and Vladimir Vernadsky.

Following the Russian Revolution of 1917 many prominent writers and philosophers left the country, including Bunin, Vladimir Nabokov and Nikolay Berdyayev, while a new generation of talented authors joined together in an effort to create a distinctive working-class culture appropriate for the new Soviet state. In the 1930s censorship over literature was tightened in line with the policy of socialist realism. In the late 1950s restrictions on literature were eased, and by the 1970s and 1980s, writers were increasingly ignoring official guidelines. Leading authors of the Soviet era include novelists Yevgeny Zamyatin, Ilf and Petrov, Mikhail Bulgakov and Mikhail Sholokhov, and poets Vladimir Mayakovsky, Yevgeny Yevtushenko, and Andrey Voznesensky.

The Soviet Union was also a major producer of science fiction, written by authors like Arkady and Boris Strugatsky, Kir Bulychov, Alexander Belayev and Ivan Yefremov.[222] Traditions of Russian science fiction and fantasy are continued today by numerous writers.

Cinema, animation and media

Former Russian President Dmitry Medvedev in the Washington studio of Russia Today TV

Russian and later Soviet cinema was a hotbed of invention in the period immediately following the 1917, resulting in world-renowned films such as *The Battleship Potemkin* by Sergei Eisenstein.[] Eisenstein was a student of filmmaker and theorist Lev Kuleshov, who developed the Soviet montage theory of film editing at the world's first film school, the All-Union Institute of Cinematography. Dziga Vertov, whose *kino-glaz* ("film-eye") theory—that the camera, like the human eye, is best used to explore real life—had a huge impact on the development of documentary film making and cinema realism. The subsequent state policy of socialist realism somewhat limited creativity, however many Soviet films in this style were artistically successful, like *Chapaev*, *The Cranes Are Flying*, and *Ballad of a Soldier*.[]

1960s and 1970s saw a greater variety of artistic styles in the Soviet cinema. Eldar Ryazanov's and Leonid Gaidai's comedies of that time were immensely popular, with many of the catch phrases still in use today. In 1961–68 Sergey Bondarchuk directed an Oscar-winning film adaptation of Leo Tolstoy's epic *War and Peace*, which was the most expensive film made in the Soviet Union.[223] In 1969, Vladimir Motyl's *White Sun of the Desert* was released, a very popular film in a genre of ostern; the film is traditionally watched by cosmonauts before any trip into space.[224]

Shukhov Tower in Moscow served for the early radio and TV broadcasting.

Russian animation dates back to the late Russian Empire times. During Soviet era, Soyuzmultfilm studio was the largest animation producer. Soviet animators developed a great variety of pioneering techniques and aesthetic styles, with prominent directors including Ivan Ivanov-Vano, Fyodor Khitruk and Aleksandr Tatarsky. Many Soviet cartoon heroes, such as the Russian-style Winnie-the-Pooh, cute little Cheburashka, Wolf and Hare from *Nu, Pogodi!* are iconic images in Russia and many surrounding countries.

The late 1980s and 1990s were a period of crisis in Russian cinema and animation. Although Russian filmmakers became free to express themselves, state subsidies were drastically reduced, resulting in fewer films produced. The early years of the 21st century have brought increased viewership and subsequent prosperity to the industry on the back of the economic revival. Production levels are already higher than in Britain and Germany.[225] Russia's total box-office revenue in 2007 was $565 million, up 37% from the previous year.[226] In 2002 the *Russian Ark* became the first feature film ever to be shot in a single take. The traditions of Soviet animation were developed recently by such directors as Aleksandr Petrov and studios like Melnitsa Animation.

Russia was among the first countries to introduce radio and television. While there were few channels in the Soviet time, in the past two decades many new state and private-owned radio stations and TV channels appeared. In 2005 a state-run English language Russia Today TV started broadcasting, and its Arabic version Rusiya Al-Yaum was launched in 2007.

Sports

Russian football fans with a gigantic "Go Russia!" banner, featuring the Russian Bear on the background of the Russian flag

Combining the total medals of Soviet Union and Russia, the country is second among all nations by number of gold medals both at the Summer Olympics and at the Winter Olympics. Soviet and later Russian athletes have always been in the top three for the number of gold medals collected at the Summer Olympics. Soviet gymnasts, track-and-field athletes, weight lifters, wrestlers, boxers, fencers, shooters, cross country skiers, biathletes, speed skaters and figure skaters were consistently among the best in the world, along with Soviet basketball, handball, volleyball and ice hockey players.[227] The 1980 Summer Olympics were held in Moscow while the 2014 Winter Olympics will be hosted in Sochi.

Although ice hockey was only introduced during the Soviet era, the national team managed to win gold at almost all the Olympics and World Championships they contested. Russian players Valery Kharlamov, Sergey Makarov, Vyacheslav Fetisov and Vladislav Tretiak hold four of six positions in the IIHF *Team of the Century*.[228] Recently Russia won the 2008, 2009 and the 2012 IIHF World Championships, overtaking Canada as the world's top ranked ice hockey team.[229] Russia totally dominated the 2012 tournament, winning all of its ten matches—the first time any team had done so since the Soviet Union in 1989.[230] The Kontinental Hockey League (KHL) was founded in 2008 as a successor to the Russian Superleague. It is seen as a rival to the National Hockey League (NHL) and is ranked the top hockey league in Europe as of 2009.[231] Bandy, also known as Russian hockey, is another traditionally popular ice sport.[232] The Soviet Union won all the Bandy World Championships between 1957–79.[233]

Dmitry Medvedev with Russia men's national ice hockey team

Along with ice hockey and basketball, association football is one of the most popular sports in modern Russia. The Soviet national team became the first ever European Champions by winning Euro 1960. In recent years, Russian football, which downgraded in 1990s, has experienced a revival. Russian clubs CSKA Moscow and Zenit St Petersburg won the UEFA Cup in 2005 and 2008 respectively. The Russian national football team reached the semi-finals of Euro 2008, losing only to the eventual champions Spain. Russia will host the 2018 FIFA World Cup, with 14 host cities located in the European part of the country and on the Urals. In 2007, the Russian national basketball team also won the European Basketball Championship. Russian basketball club PBC CSKA Moscow is one of the top teams in Europe, winning the Euroleague in 2006 and 2008.

Larisa Latynina, who currently holds a record for most Olympic medals won per person and most gold Olympic medals won by a woman, established the USSR as the dominant force in gymnastics for many years to come.[234] Today, Russia is leading in rhythmic gymnastics with Alina Kabayeva, Irina Tschaschina and Yevgeniya Kanayeva. Russian synchronized swimming is the best in the world, with almost all gold medals at Olympics and World Championships having been swept by Russians in recent decades. Figure skating is another popular sport in Russia, especially pair skating and ice dancing. At every Winter Olympics from 1964 until 2006 a Soviet or Russian pair has won gold. Since the end of the Soviet era, tennis has grown in popularity and Russia has produced a number of famous players, including Maria Sharapova, the world's highest paid female athlete.[235] In martial arts, Russia produced the sport Sambo and many renown fighters, like Fedor Emelianenko. Chess is a widely popular pastime in Russia; from 1927, Russian grandmasters have held the world chess championship almost continuously.[236]

Formula One is also becoming increasingly popular in Russia. Renault's Vitaly Petrov is the only Russian Formula One driver to date. There have only ever been two Russian Grands Prix (in 1913 and 1914), but it is set to return for 2014, in a six-year deal.[237]

National holidays and symbols

2005 Victory Day (9 May) parade on Moscow's Red Square

There are seven public holidays in Russia,[238] except those always celebrated on Sunday. Russian New Year traditions resemble those of the Western Christmas, with New Year Trees and gifts, and Ded Moroz (Father Frost) playing the same role as Santa Claus. Orthodox Christmas falls on 7 January, because Russian Orthodox Church still follows the Julian calendar and all Orthodox holidays are 13 days after Western ones. Another two major Christian holidays are Easter and Trinity Sunday. Kurban Bayram and Uraza Bayram are celebrated by Russian Muslims.

Further Russian public holidays include Defender of the Fatherland Day (23 February), which honors Russian men, especially those serving in the army; International Women's Day (8 March), which combines the traditions of Mother's Day and Valentine's Day; Spring and Labor Day (1 May); Victory Day (9 May); Russia Day (12 June); and Unity Day (4 November), commemorating the popular uprising which expelled the Polish occupation force from Moscow in 1612.

Victory Day is the second most popular holiday in Russia; it commemorates the victory over Nazism in the Great Patriotic War. A huge military parade, hosted by the President of Russia, is annually organised in Moscow on Red Square. Similar parades took place in all major Russian cities and cities with the status *Hero city* or *City of Military Glory*.

Popular non-public holidays include Old New Year (New Year according to Julian Calendar on 14 January), Tatiana Day (students holiday on 25 January), Maslenitsa (a pre-Christian spring holiday a week before the Great Lent), Cosmonautics Day (in tribute to the first human trip into space), Ivan Kupala Day (another pre-Christian holiday on 7 July) and Peter and Fevronia Day (taking place on 8 July and being the Russian analogue of Valentine's Day, which focuses, however, on the family love and fidelity).

Scarlet Sails celebration on the Neva river in Saint Petersburg

State symbols of Russia include the Byzantine double-headed eagle, combined with St. George of Moscow in the Russian coat of arms. The Russian flag dates from the late Tsardom of Russia period and has been widely used since the time of the Russian Empire. The Russian anthem shares its music with the Soviet Anthem, though not the lyrics. The imperial motto *God is with us* and the Soviet motto *Proletarians of all countries, unite!* are now obsolete and no new motto has replaced them. The hammer and sickle and the full Soviet coat of arms are still widely seen in Russian cities as a part of old architectural decorations. The Soviet Red Stars are also encountered, often on military equipment and war memorials. The Red Banner continues to be honored, especially the Banner of Victory of 1945.

The Matryoshka doll is a recognizable symbol of Russia, and the towers of Moscow Kremlin and Saint Basil's Cathedral in Moscow are main Russia's architectural icons. Cheburashka is a mascot of the Russian national Olympic team. St. Mary, St. Nicholas, St. Andrew, St. George, St. Alexander Nevsky, St. Sergius of Radonezh and St. Seraphim of Sarov are Russia's patron saints. Chamomile is the national flower, while birch the national tree. The Russian bear is an animal symbol and a national personification of Russia, though this image has a Western origin and Russians themselves have accepted it only fairly recently. The native Russian national personification is Mother Russia.

Tourism

Grand Cascade in Peterhof, a popular tourist destination in Saint Petersburg

Tourism in Russia has seen rapid growth since the late Soviet times, first domestic tourism and then international tourism, fueled by the rich cultural heritage and great natural variety of the country. Major tourist routes in Russia include a journey around the Golden Ring of ancient cities, cruises on the big rivers like the Volga, and long journeys on the famous Trans-Siberian Railway.

The most visited destinations in Russia are Moscow and Saint Petersburg, the current and the former capitals of the country. Recognized as World Cities, they feature such world-renown museums as Tretyakov Gallery and Hermitage, famous theaters like Bolshoi and Mariinsky, ornate churches like Saint Basil's Cathedral, Cathedral of Christ the Saviour, Saint Isaac's Cathedral and Church of the Savior on Blood, impressive fortifications like Moscow Kremlin and Peter and Paul Fortress, beautiful squares and streets like Red Square, Palace Square, Tverskaya Street and Nevsky Prospect. Rich palaces and parks are found in the former imperial residences in suburbs of Moscow (Kolomenskoye, Tsaritsyno) and St Petersburg (Peterhof, Strelna, Oranienbaum, Gatchina, Pavlovsk and Tsarskoye Selo). Moscow displays the Soviet architecture at its best, along with modern skyscrapers, while St Petersburg, nicknamed *Venice of the North*, boasts of its classical architecture, many rivers, channels and bridges.

Matryoshka doll taken apart

Kazan, the capital of Tatarstan, shows a mix of Christian Russian and Muslim Tatar cultures. The city has registered a brand *The Third Capital of Russia*, though a number of other major cities compete for this status, including Novosibirsk, Yekaterinburg and Nizhny Novgorod.

Typical Russian souvenirs include matryoshka doll and other handicrafts, samovars for water heating, ushanka and papaha warm hats, and fur clothes. Russian vodka and caviar are among the food that attracts foreigners.

The warm subtropical Black Sea coast of Russia is the site for a number of popular sea resorts, like Sochi, the follow-up host of the 2014 Winter Olympics. The mountains of the Northern Caucasus contain popular ski resorts, including Dombay. The most famous natural destination in Russia is Lake Baikal, *the Blue Eye of Siberia*. This unique lake, oldest and deepest in the world,[] has crystal-clean waters and is surrounded by taiga-covered mountains. Other popular natural destinations include Kamchatka with its volcanoes and geysers, Karelia with its lakes and granite rocks, the snowy Altai Mountains, and the wild steppes of Tyva.

See also

- Culture of Russia
- Geology of Russia
- Index of Soviet Union-related articles
- International rankings of Russia
- Outline of Russia
- Timeline of Russian history

References

[1] http://www.gks.ru/bgd/free/B13_00/IssWWW.exe/Stg/dk04/8-0.htm

[2] http://data.worldbank.org/indicator/NY.GDP.MKTP.PP.CD?order=wbapi_data_value_2012+wbapi_data_value+wbapi_data_value-last&sort=desc

[3] http://data.worldbank.org/indicator/NY.GDP.PCAP.PP.CD?order=wbapi_data_value_2012+wbapi_data_value+wbapi_data_value-last&sort=desc

[4] http://data.worldbank.org/indicator/NY.GDP.MKTP.CD

[5] hhttp://data.worldbank.org/indicator/NY.GDP.PCAP.CD/countries/1W?display=default

[7] "The names Russian Federation and Russia shall be equal".

[8] (Russian) Official estimate (http://www.gks.ru/bgd/free/B12_00/IssWWW.exe/Stg/dk10/8-0.htm) as of 1 September 2012.

[9] *The Mongol empire: its rise and legacy*, By Michael Prawdin, Gérard Chaliand, (2005) page 512-550

[11] Peter Turchin, Thomas D. Hall and Jonathan M. Adams, " East-West Orientation of Historical Empires (http://jwsr.ucr.edu/archive/vol12/number2/pdf/jwsr-v12n2-tah.pdf)", *Journal of World-Systems Research* Vol. 12 (no. 2), pp. 219–229 (2006).

[14] Rozhnov, Konstantin, Who won World War II? (http://news.bbc.co.uk/2/hi/europe/4508901.stm). BBC.

[16] International Energy Agency – Oil Market Report 18 January 2012 (http://omrpublic.iea.org/omrarchive/18jan12sup.pdf). Retrieved 20 February 2012.

[17] Country Comparison :: Natural gas – production (https://www.cia.gov/library/publications/the-world-factbook/rankorder/2180rank.html). CIA World Factbook. Retrieved 17 February 2012.

[38] The history of banya and sauna (http://web.archive.org/web/20121130084214/http://sauna-banya.ru/ist.html) (Russian)

[39] " *Black Death* (http://books.google.com/books?id=yw3HmjRvVQMC&pg=PA62)". Joseph Patrick Byrne (2004). p. 62. ISBN 0-313-32492-1

[41] Frank D. McConnell. Storytelling and Mythmaking: Images from Film and Literature. (http://books.google.com/books?id=rqhZAAAAMAAJ) Oxford University Press, 1979. ISBN 0-19-502572-5; p. 78: "But Ivan IV, Ivan the Terrible, or as the Russian has it, *Ivan Groznyi*, "Ivan the Magnificent" or "Ivan the Awesome," is precisely a man who has become a legend"

[52] Geoffrey A. Hosking (2001). " *Russia and the Russians: a history* (http://books.google.com/books?id=oh-5AAmboMUC&pg=PA9)". Harvard University Press. p. 9. ISBN 0-674-00473-6

[53] N. M. Dronin, E. G. Bellinger (2005). " *Climate dependence and food problems in Russia, 1900–1990: the interaction of climate and agricultural policy and their effect on food problems* (http://books.google.com/books?id=9a5j_JL6cqIC&pg=PA38)". Central European University Press. p. 38. ISBN 963-7326-10-3

[54] Transactions of the American Philosophical Society (http://books.google.com/books?id=uUsLAAAAIAAJ&pg=PA3). James E. Hassell (1991), p. 3. ISBN 0-87169-817-X

[55] Famine in Russia: the hidden horrors of 1921 (http://web.archive.org/web/20121219101057/http://www.icrc.org/eng/resources/documents/misc/5rfhjy.htm), International Committee of the Red Cross

[56] Abbott Gleason (2009). *A Companion to Russian History* (http://books.google.com/books?id=JyN0hlKcfTcC&pg=PA373). Wiley-Blackwell. p. 373. ISBN 1-4051-3560-3

[57] Getty, Rittersporn, Zemskov. Victims of the Soviet Penal System in the Pre-War Years: A First Approach on the Basis of Archival Evidence. The American Historical Review, Vol. 98, No. 4 (Oct. 1993), pp. 1017–49.

[58] R.W. Davies, S.G. Wheatcroft (2004). *The Years of Hunger: Soviet Agriculture, 1931–33*, pp. 401.

[61] The Legacy of the Siege of Leningrad, 1941–1995 (http://www.cambridge.org/gb/knowledge/isbn/item1173696/?site_locale=en_GB). Cambridge University Press.

[63] Geoffrey A. Hosking (2006). " *Rulers and victims: the Russians in the Soviet Union* (http://books.google.com/books?id=CDMVMqDvp4QC&pg=PA242)". Harvard University Press. p. 242. ISBN 0-674-02178-9

[65] Foreign trade (http://rs6.loc.gov/cgi-bin/query/r?frd/cstdy:@field(DOCID+su0391)) from *A Country Study: Soviet Union (Former)*. Library of Congress Country Studies project.

[80] Venezuela's President Hugo Chavez recognizes independence of breakaway Georgia republics (http://articles.latimes.com/2009/sep/11/world/fg-russia-chavez11) by Megan K. Stack. 9 September 2009

[81] Netanyahu declares Russia as superpower (http://freevideo.rt.com/video/1759) Russia Today News 15 February 2010

[82] Superpower Reborn (http://georgiandaily.com/index.php?option=com_content&task=view&id=6527) by Ronald Steel. New York Times, 24 August 2008

[83] Russia is a Superpower CNN, US Senators telling the truth (http://video.google.com/videoplay?docid=-9079543725663390621) CNN News 30 August 2008

[84] Steven Rosefielde, Russia in the 21st Century The Prodigal Superpower (http://books.google.com/books?id=eC6HdSYZhRgC), Cambridge University Press, 2005 ISBN 0521836786

[85] Is Russia a Superpower? Cold War II? (http://atlanticreview.org/archives/1160-Is-Russia-a-Superpower-Cold-War-II.html) Atlantic Review, 25 August 2008.

[86] What's Looming in Ukraine Is more Threatening than Georgia (http://www.spiegel.de/international/world/0,1518,584631,00.html) Der Spiegel, 16 October 2008. Quote: "Nikonov: Russia is not a superpower and won't be one for the foreseeable future. But Russia is a great power. It was one, it is one and it will continue to be one."

[88] Human Rights Watch on Russia and Chechnya HTW.org (http://www.hrw.org/en/video/2008/04/06/russia-trial)

[91] The Shanghai Cooperation Organisation (http://www.globalsecurity.org/military/world/int/sco.htm) at Globalsecurity.org 27 April 2005

[92] (Russian)

[95] Russia pilots proud of flights to foreign shores (http://www.komonews.com/news/national/28390779.html) by David Nowak. The Associated Press, 15 September 2008

[103] Direct elections of heads of republics (and other federal subjects) are mandated by art. 18 of the Federal Law of 06.10.1999 № 184-FZ (http://constitution.garant.ru/act/federative/12117177/chapter/3/#300) as amended by Federal Law of 2.05.2012 № 40-FZ (http://base.garant.ru/70169404/#131)

[104] Designation of republican heads as presidents is forbidden by Federal Law of 28.12.2010 № 406-FZ (http://graph.document.kremlin.ru/page.aspx?1;1538006), but transitional period lasts until 1 January 2015

[105] *Russian Classification of Economic Regions* (OK 024–95) of 1 January 1997 as amended by the Amendments #1/1998 through #5/2001. (Section I. Federal Districts)

[122] Debt – external (https://www.cia.gov/library/publications/the-world-factbook/rankorder/2079rank.html), CIA World Factbook. Retrieved 22 May 2010.

[127] GRP by federal subjects of Russia, 1998–2007 (http://www.gks.ru/bgd/free/b01_19/IssWWW.exe/Stg/d000/vrp98-07.htm) (Russian)

[131] Land Use (https://www.cia.gov/library/publications/the-world-factbook/fields/2097.html) CIA World Factbook

[132] Data (http://www.gks.ru/bgd/regl/b09_02/IssWWW.exe/Stg/d010/1-03.doc) by Rosstat (Russian)

[133] Russia takes the third place in the world by grain exports (http://web.archive.org/web/20120108052630/http://www.rosbankjournal.ru/news/11588), rosbankjournal.ru (Russian)

[134] Data (http://www.gks.ru/bgd/regl/b09_02/IssWWW.exe/Stg/d010/1-04.doc) by Rosstat (Russian)

[135] Agricultural land by type of owners (http://www.gks.ru/bgd/regl/b09_38/IssWWW.exe/Stg/d01/04-01.htm) Rosstat, 2009 (Russian)

[136] Main agricultural products by type of owners (http://www.gks.ru/bgd/regl/b09_38/IssWWW.exe/Stg/d01/02-28.htm) Rosstat, 2009 (Russian)

[138] Exports and imports of fish and sea products (http://www.gks.ru/bgd/regl/b09_13/IssWWW.exe/Stg/html4/15-07.htm) Rosstat, 2009 (Russian)

[139] FAO. 2010. Global Forest Resources Assesment 2010. Main Report. FAO Forestry Working Paper 163, Rome, Italy (http://www.fao.org/docrep/014/i1757r/i1757r.pdf) (Russian)

[142] FAO. 2012. Russian Federation Forest Sector Outlook Study to 2030. Rome, Italy (http://www.fao.org/docrep/016/i3020r/i3020r00.pdf) (Russian)

[143] Russia, China in Deal On Refinery, Not Gas (http://online.wsj.com/article/SB10001424052748704129204575505373365685564.html) by Jacob Gronholt-Pedersen. Wall Street Journal, 22 September 2010

[144] Did A New Pipeline Just Make Russia The Most Important Energy Superpower By Far (http://www.businessinsider.com/why-russia-is-about-to-change-the-world-2010-1) by Graham Winfrey. Business Insider, 6 January 2010

[145] Country Comparison :: Natural gas – proved reserves (https://www.cia.gov/library/publications/the-world-factbook/rankorder/2179rank.html). CIA World Factbook. Retrieved 5 October 2010.

[146] Country Comparison :: Oil – proved reserves (https://www.cia.gov/library/publications/the-world-factbook/rankorder/2178rank.html). CIA World Factbook. Retrieved 5 October 2010.

[148] Country Comparison :: Natural gas – exports (https://www.cia.gov/library/publications/the-world-factbook/rankorder/2183rank.html). CIA World Factbook. Retrieved 5 October 2010.

[150] Country Comparison :: Electricity – production (https://www.cia.gov/library/publications/the-world-factbook/rankorder/2038rank.html). CIA World Factbook. Retrieved 5 October 2010.

[152] Nuclear Power Plant Information (http://www.iaea.org/programmes/a2/index.html), International Atomic Energy Agency. Retrieved 12 June 2006.

[153] Russia builds nuclear power stations all over the world (http://amur.kp.ru/daily/24458/620649/) at amur.kp.ru (Russian)

[155] CIS railway timetable (http://www.poezda.net/en/web), route No. 002, Moscow-Pyongyang, August 2009. Note: several different routes have the same number.
[156] CIS railway timetable (http://www.poezda.net/en/web), route No. 350, Kiev-Vladivostok, August 2009.
[157] Rosstat statistics on length of roads (http://www.gks.ru/bgd/regl/b08_11/IssWWW.exe/Stg/d02/18-09.htm) Retrieved 10 June 2009
[159] http://englishrussia.com/index.php/2008/03/31/russian-atomic-icebreakers/
[164] Panzerkampfwagen T-34(r) (http://www.achtungpanzer.com/panzerkampfwagen-t-34r-soviet-t-34-in-german-service.htm) by George Parada (n.d.) *Achtung Panzer!* website. Retrieved 17 November 2008
[165] Halberstadt, Hans Inside the Great Tanks The Crowood Press Ltd. Wiltshire, England 1997 94–96 ISBN 1-86126-270-1: "The T-54/T-55 series is the hands down, all time most popular tank in history."
[167] Medvedev outlines priorities for Russian economy's modernization (http://en.rian.ru/russia/20091011/156428675.html) RIA Novosti
[170] ОБ ИТОГАХ ВСЕРОССИЙСКОЙ ПЕРЕПИСИ НАСЕЛЕНИЯ 2010 ГОДА (*Results of the Russian 2010 census*) (http://www.perepis-2010.ru/results_of_the_census/result-december-2011.ppt) – http://www.perepis-2010.ru/
[171] Ethnic groups in Russia (http://demoscope.ru/weekly/ssp/rus_nac_02.php), 2002 census, Demoscope Weekly. Retrieved 5 February 2009.
[174] Modern demographics of Russia (http://www.gks.ru/free_doc/2010/demo/dem-sit-09.doc) by Rosstat. Retrieved on 5 October 2010
[176] Putin tries to lure millions of Russian expats home (http://www.timesonline.co.uk/tol/news/world/europe/article728982.ece) Times Online. 9 February 2006.
[177] Moya Flynn. Migrant resettlement in the Russian federation: reconstructing 'homes' and 'homelands' (http://books.google.com/books?id=YLeAxHLmgR8C&pg=PA15), Anthem Press (2004). p. 15. ISBN 1-84331-117-8
[182] Russian birth rates 1950–2008 (http://demoscope.ru/weekly/ias/ias05.php?tim=0&cou=26&terr=1&ind=26&Submit=OK) Demoscope Weekly. Retrieved October 2010.
[192] (registration required)
[200] Russian life expectancy figures (http://www.gks.ru/free_doc/new_site/population/demo/demo26.xls) Rosstat. Retrieved 21 August 2010
[201] David Johnson, ed., *Politics, Modernisation and Educational Reform in Russia: From Past to Present* (2010)
[207] The first stone tented roof church and the origins of the tented roof architecture (http://web.archive.org/web/20130116214934/http://www.rusarch.ru/zagraevsky19.htm) by Sergey Zagraevsky at RusArch.ru **(Russian)**
[208] The shapes of domes of ancient Russian churches (http://web.archive.org/web/20130116215900/http://www.rusarch.ru/zagraevsky1.htm) by Sergey Zagraevsky at the site of RusArch.ru **(Russian)**
[209] Russian: Постановление ЦК КПСС и СМ СССР "Об устранении излишеств в проектировании и строительстве", 04.11.1955 (Khrushchev's decree *On liquidation of excesses...*) **(Russian)**
[210] Over 20,000 churches rebuilt in Russia in 20 years – Patriarch Kirill (http://www.en.rian.ru/art_living/20101203/161616486.html) RIA Novosti
[211] Russian Academy of Arts (http://en.rah.ru/content/en/section-academy-en.html) official site.
[218] History of Rock Music in Russia (http://www.russia-ic.com/culture_art/music/380/#r1) at Russia-InfoCentre
[223] Birgit Beumers. *A History of Russian Cinema*. Berg Publishers (2009). ISBN 978-1-84520-215-6. p. 143.

External links

- Russia (http://www.dmoz.org/Regional/Europe/Russia/) at the Open Directory Project
- Wikimedia Atlas of Russia
- Russia (https://www.cia.gov/library/publications/the-world-factbook/geos/rs.html) entry at *The World Factbook*
- Russia (http://ucblibraries.colorado.edu/govpubs/for/russia.htm) at *UCB Libraries GovPubs*
- Russia (http://www.bbc.co.uk/news/world-europe-17839672) from the BBC News
- Russia (http://www.britannica.com/EBchecked/topic/513251/Russia) at *Encyclopædia Britannica*

Government

- Official Russian governmental portal (http://www.gov.ru/index_en.html)
- Chief of State and Cabinet Members (https://www.cia.gov/library/publications/world-leaders-1/world-leaders-r/russia.html)
- Russian News Agency "Ria Novosti" (http://en.rian.ru/)
- Russian radio "Voice of Russia" (http://english.ruvr.ru/2010/09/24/22168374.html)

Other

- Post-Soviet Problems (http://repository.library.georgetown.edu/handle/10822/552653) from the Dean Peter Krogh Foreign Affairs Digital Archives (http://repository.library.georgetown.edu/handle/10822/552494/browse?type=title)
- Russia Beyond the Headlines (http://rbth.ru/) International news project about Russia
- Way to Russia. An Introduction to Russia and Russian People (http://www.waytorussia.net/WhatIsRussia/Intro.html)
- Russia cities and regions guide (http://russiatrek.org/)
- Official Russia Travel Guide (http://www.russia-travel.com/)
- Russian Consulate (http://russianconsulate.com/)
- Russia Beyond the Headlines (http://rbth.ru/) International news project about Russia
- Moscow Russia Insider's Guide (http://www.moscow-russia-insiders-guide.com/) Moscow and Russia through Muscovite's eyes.

Krai

Krai or **kray** (Russian: край) was a type of an administrative division in the Russian Empire (see Krais of the Russian Empire) and the Russian SFSR, and is one of the types of the federal subjects of modern Russia (see Krais of Russia).

Etymologically, the word is related to the verb "кроить" (*kroit'*), i.e., "to cut".[1] Historically, krais were vast territories located along the periphery of Russia, since the word *krai* also means *border* or *edge*, i.e., *a place of the cut-off*. Currently the administrative usage of the term is mostly traditional, as some oblasts also fit this description and there is no difference in legal status between the krais and the oblasts.[2]

See also

- Kraj, an equivalent term used in the Czech Republic and Slovakia
- Krajina
- Kreis, an administrative unit in Germany
- Marches (Mark, in German) compare to Denmark (literally - the borderland of Danes).
- Ukraine

References

[1] Etymology of the word (http://starling.rinet.ru/cgi-bin/response.cgi?root=/usr/local/share/starling/morpho&morpho=1&basename=\usr\local\share\starling\morpho\vasmer\vasmer&first=1&text_word=Ð°ÑÐ°Ð¹&method_word=substring&text_general=&method_general=substring&text_origin=&method_origin=substring&text_trubachev=&method_trubachev=substring&text_editorial=&method_editorial=substring&text_pages=&method_pages=substring&text_any=&method_any=substring&sort=word) (**Russian**)

[2] 12 декабря 1993 г «Конституция Российской Федерации. Статья 5.», в ред. Федерального конституционного закона №5-ФКЗ от 21 июля 2007 г. (December 12, 1993 *Constitution of the Russian Federation. Article 5.*, as amended by the Federal Constitutional Law #5-FKZ of July 21, 2007.).

Subdivisions_of_Russia

Russia is divided into several **types and levels of subdivisions**.

Federal subjects

Russia is a federation which since March 1, 2008 consists of 83 federal subjects (members of the Federation).[1] These federal subjects are of equal federal rights in the sense that they have equal representation—two delegates each—in the Federation Council (upper house of the Federal Assembly). They do, however, differ in the degree of autonomy they enjoy.

Federal subjects of Russia

Six types of federal subjects are distinguished—21 republics, 9 krais, 46 oblasts, 2 federal cities, 1 autonomous oblast, and 4 autonomous okrugs.

Autonomous okrugs are the only ones that have a peculiar status of being federal subjects in their own right, yet at the same time they are considered to be administrative divisions of other federal subjects (with Chukotka Autonomous Okrug being the only exception).

Administrative divisions

Prior to the adoption of the 1993 Constitution of Russia, the administrative-territorial structure of Russia was regulated by the Decree of the Presidium of the Supreme Soviet of the RSFSR of August 17, 1982 "On the Procedures of Dealing with the Matters of the Administrative-Territorial Structure of the RSFSR".[2] The 1993 Constitution, however, did not identify the matters of the administrative-territorial divisions as the responsibility of the federal government or as the joint responsibility of the federal government and the federal subjects.[2] This was interpreted by the governments of the federal subjects as a sign that the matters of the administrative-territorial divisions became solely the responsibility of the federal subjects.[2] As a result, the modern administrative-territorial structures of the federal subjects vary significantly from one federal subject to another. While the implementation details may be considerably different, in general, however, the following types of high-level administrative divisions are recognized:

- raions (administrative districts)
- cities/towns and urban-type settlements of federal subject significance
- closed administrative-territorial formations

Autonomous okrugs and okrugs are intermediary units of administrative divisions, which include some of the federal subject's raions and cities/towns/urban-type settlements of the federal subject significance.

- Autonomous okrugs, while being under the jurisdiction of another federal subject, are still constitutionally recognized as federal subjects on their own right. Chukotka Autonomous Okrug is an exception to this description in that it is not administratively subordinated to any other federal subject of Russia.

- Okrugs are usually former autonomous okrugs which lost their federal subject status due to mergers with other federal subjects.

Typical lower-level administrative divisions include:

- selsoviets (rural councils)
- towns and urban-type settlements of the administrative district significance
- city districts

Municipal divisions

In the course of the Russian municipal reform of 2004–2005, all federal subjects of Russia were to streamline the structures of local self-government, which is guaranteed by the Constitution of Russia. The reform mandated that each federal subject was to have a unified structure of the municipal government bodies by January 1, 2005, and a law enforcing the reform provisions went in effect on January 1, 2006. According to the law, the units of the municipal division (called "**municipal formations**") are as follows:[3]

- **Municipal district**, a group of urban and rural settlements, often along with the inter-settlement territories. In practice, municipal districts are usually formed within the boundaries of existing administrative districts (*raions*).
  - **Urban settlement**, a city/town or an urban-type settlement, possibly together with adjacent rural and/or urban localities
  - **Rural settlement**, one or several rural localities
- **Urban okrug**, an urban settlement not incorporated into a municipal district. In practice, urban okrugs are usually formed within the boundaries of existing cities of federal subject significance.
- **Intra-city territory of a federal city**, a part of a federal city's territory. In Moscow, these are called **municipal formations** (which correspond to districts); in St. Petersburg—*municipal okrugs*, *towns*, and *settlements*.

Territories not included as a part of municipal formations are known as *inter-settlement territories*.

Other types of subdivisions

Federal districts

Federal districts of Russia

All of the federal subjects are grouped into eight **federal districts**,[4] each administered by an envoy appointed by the President of Russia. Federal districts' envoys serve as liaisons between the federal subjects and the federal government and are primarily responsible for overseeing the compliance of the federal subjects with the federal laws.

Economic regions

Economic regions of Russia

For economic and statistical purposes the federal subjects are grouped into twelve **economic regions**.[5] Economic regions and their parts sharing common economic trends are in turn grouped into economic zones and macrozones.

See also

- History of the administrative division of Russia
- List of federal subjects of Russia by area
- List of federal subjects of Russia by population
- Types of inhabited localities in Russia

References

[1] Constitution, Article 65

[2] "Энциклопедический словарь конституционного права". Статья "Административно-территориальное устройство". Сост. А. А. Избранов. — Мн.: Изд. В.М. Суров, 2001.

[3] Государственная Дума Российской Федерации. Федеральный Закон №131-ФЗ от 6 октября 2003 г. «Об общих принципах организации местного самоуправления в Российской Федерации», в ред. Федерального Закона №243-ФЗ от 28 сентября 2010 г. (State Duma of the Russian Federation. Federal Law #131-FZ of October 6, 2003 *On General Principles of the Organization of Local Self-Government in the Russian Federation*, as amended by the Federal Law #243-FZ of September 28, 2010.).

[4] Президент Российской Федерации. Указ №849 от 13 мая 2000 г. «О полномочном представителе Президента Российской Федерации в федеральном округе». Вступил в силу 13 мая 2000 г. Опубликован: "Собрание законодательства РФ", №20, ст. 2112, 15 мая 2000 г. (President of the Russian Federation. Decree #849 of May 13, 2000 *On the Plenipotentiary Representative of the President of the Russian Federation in a Federal District*. Effective as of May 13, 2000.).

[5] "Общероссийский классификатор экономических регионов" (ОК 024-95) введённый 1 января 1997 г., в ред. Изменения № 05/2001. Секция II. Экономические районы (*Russian Classification of Economic Regions* (OK 024-95) of January 1, 1997 as amended by the Amendments #1/1998 through #5/2001. Section II. Economic Regions)

Sources

- 12 декабря 1993 г. « Конституция Российской Федерации (http://www.consultant.ru/online/base/?req=doc;base=LAW;n=83686)», в ред. Федерального конституционного закона №7-ФКЗ от 30 декабря 2008 г. Вступил в силу со дня официального опубликования. Опубликован: "Российская газета", №237, 25 декабря 1993 г. (December 12, 1993 *Constitution of the Russian Federation* (http://translate.google.com/translate?hl=en&sl=ru&tl=en&u=http://www.consultant.ru/online/base/?req=doc;base=LAW;n=83686), as amended by the Federal Constitutional Law #7-FKZ of December 30, 2008. Effective as of the official publication date.).

External links

- *Local government in Russia: Its powers vary across the country* at Citymayors.com (http://www.citymayors.com/government/russia_government.html) **(English)**

Administrative_centre

An **administrative centre** is a seat of regional administration or local government, or a county town, or the place where the central administration of a commune is located.

In Russia, the term is applied to the inhabited localities which serve as a seat of government of entities of various levels. The only exception to this rule is the republics, for which the term "capital" is used to refer to the seat of government. The capital of Russia is also an entity to which the term "administrative centre" does not apply. A similar arrangement exists in Ukraine.

In the United Kingdom it is the centre of a local authority which is distinct from a historic county with a county town.

See also

- Local government
- Chef-lieu, administrative centres in Algeria, Belgium, Luxembourg, France, Switzerland, and Tunisia
- County seat, administrative centres in the USA
- County town, administrative centres in Ireland and the UK
- List of regions of Norway
- Administrative divisions of Portugal
- Subdivisions of Russia

References

Russian_Census_(2002)

Russian Census of 2002 (Russian: Всеросси́йская пе́репись населе́ния 2002 го́да) was the first census of the Russian Federation carried out on October 9 through October 16, 2002. It was carried out by the Russian Federal Service of State Statistics (Rosstat).

Data collection

The census data were collected as of midnight October 9, 2002.

Resident population

The census was primarily intended to collect statistical information about the **resident population** of Russian Federation. The resident population included:

- Russian citizens living in Russia (including those temporary away from the country, provided the absence from the country was expected to last less than one year);
- non-citizens (i.e. foreign citizens and stateless persons) who were any of the following:
  - legal permanent residents;
  - persons who have arrived to the country with the intent to settle permanently or to seek asylum, regardless of whether they have actually obtained the appropriate immigration status;
  - authorized foreign workers or students, provided the period of temporary residence in Russia was expected to last at least one year.

All detailed census tables are for the resident population.

All (resident) participants were asked question on their gender, birth date, marital status, household composition, bithplace, citizenship, ethnic or tribal self-identification (национальность), education level, language competence, sources of income, and employment status. A sample of the participants were also asked more detailed questions about their economic and housing situation.

Non-residents

In additional to the resident population, the census also counted two more groups of people:

- Russian citizens currently living abroad for more than one year in connection with the employment with the federal government.
- Persons (regardles of their citizenship) permanently residing abroad, but temporarily present in Russia (as tourists, short-term foreign workers or students, etc.).

Foreign citizens present in Russia as employees of foreign diplomatic missions or international organizations, and members of their household, were excluded from the census altogether.

Census results

The Census recorded the resident population of 145,166,731 persons, including 67,605,133 men and 77,561,598 women. That included urban population of 106,429,000 (73%) and rural population of 38,738,000 (27%).

The non-resident populations included:

- Russian citizens living abroad in connection with the federal government service: 107,288 (67,058 men and 40,230 women);
- Foreign residents present in Russia on the census date: 239,018 (177,465 men and 61,553 women).

Citizenship

Census participants were asked what country (or countries) they were citizens of. 142,442,000 respondents reported being Russian citizens; among them, 44,000 also had citizenship of another country.

Among Russia's resident population, 1,025,413 foreign citizens and 429,881 stateless persons were counted. [1]

Foreign population in Russia in 2002

	Citizens of...	
CIS countries		**906,314**
	Armenia	136,841
	Azerbaijan	154,911
	Belarus	40,330
	Georgia	52,918
	Kazakhstan	69,472
	Kyrgyzstan	28,843
	Moldova	50,988
	Tajikistan	64,165
	Turkmenistan	6,417
	Ukraine	230,558
	Uzbekistan	70,871
Other countries		**119,099**
	Afghanistan	8,221
	Bulgaria	2,262
	China	30,598
	Estonia	1,066
	Finland	285
	Germany	1,329
	Greece	612
	India	5,351
	Israel	1,016
	Latvia	2,864
	Lithuania	4,583
	Syria	1,230
	Turkey	4,991
	USA	1,361
	Vietnam	22,545
Stateless persons		429,881

1,269,023 persons did not report their citizenship.

Language abilities

Among the questions asked were "Are you competent in the Russian language?" (Владеете ли Вы русским языком?) and "What other languages are you competent in?" (Какими иными языками Вы владеете?). As the census manual explained, "competence" (владение) meant either the ability to speak, read and write a language, or only the ability to speak it. The questions did not distinguish native and non-native speakers, nor did they try to measure the degree of language competence. For small children, presumably, the recorded answer was based on the language(s) spoken by the parents.

142.6 million (98.3%) of the responders claimed competence in Russian. Other widely reported languages (more than 500,000 speakers each) are listed in the table below.

Language	Speakers (millions)
English	6.95
(Volga) Tatar	5.35
German	2.90
Ukrainian	1.81
Chechen	1.33
Chuvash	1.33
Armenian	0.91
Avar	0.78
French	0.71
Azeri	0.67
Mordvin (Moksha or Erzya)	0.61
Kabardian or Circassian	0.59
Kazakh	0.56
Dargin	0.50

1.42 million responders did not provide language information.

For a more detailed list, see List of languages of Russia.

See also

- Demographics of Russia (includes ethnic composition and some other census results)
- Russian Empire Census
- Soviet Census

External links

- Official census home page [2]
- (**Russian**) Population of Chechnya: was the Census correct? [3]

References

[1] http://www.perepis2002.ru/ct/doc/English/4-5.xls
[2] http://www.perepis2002.ru/index.html?id=87
[3] http://www.demoscope.ru/acrobat/ps96.pdf

Russian_Census_(2010)

The **Russian Census of 2010** (Russian: Всеросси́йская пе́репись населе́ния 2010 го́да) is the first census of the Russian Federation population since 2002 and the second in its post-Soviet history. Preparations for the census began in 2007 and it took place between October 14 and October 25.[1]

Inside a census station in Severodvinsk

A 10-ruble coin commemorating the 2010 Census

The census

The census was originally scheduled for October 2010, but was moved to 2013 allegedly for financial reasons,[2] although it was also speculated that political motives were influential in the decision. However in late 2009 Russian Prime Minister Vladimir Putin announced that the Government of Russia allocated 10.5 billion rubles in order to conduct the census as originally scheduled (in October 2010).[3] [4]

In the census, many Russians in Siberia entered “Siberian" as their nationality, which prompted Michael Bohm of The Moscow Times to call for removal of 'nationality' from Russian passports.[] 6,000 individuals identified themselves as Siberian in the census.[5]

Preliminary results

Preliminary results released in April 2011 showed the population to stand at 142,905,200,[6] a decline of 2.6% since the 2002 census.

See also

- Demographics of Russia

References

[1] "Осенний пересчет" (http://www.rg.ru/2010/10/15/perepis.html). "Rossiyskaya Gazeta", October 15, 2010
[2] Всероссийская перепись населения переносится на 2013 год (http://www.regnum.ru/news/1206699.html) (Russian)
[3] Путин снова перенес Всероссийскую перепись населения на 2010 год? (http://www.regnum.ru/news/1220355.html) (Russian)
[4] О ВСЕРОССИЙСКОЙ ПЕРЕПИСИ НАСЕЛЕНИЯ 2010 ГО (http://www.gks.ru/perepis/2010/new-perep10.htm) Rosstat Retrieved on November 5, 2009 (Russian)
[5] http://ajw.asahi.com/article/globe/feature/trans-siberian/AJ201209090012
[6] 2010 Russian Census (http://www.perepis-2010.ru/message-rosstat.php) Retrieved on May 25, 2011

External links

- Official website of the 2010 Census (http://www.perepis-2010.ru/) (Russian)

Chuvash_people

The **Chuvash people** (Chuvash: Чăвашсем; Russian: Чуваши) are a Turkic ethnic group, native to an area stretching from the Volga Region to Siberia. Most of them live in Republic of Chuvashia and surrounding areas, although Chuvash communities may be found throughout the Russian Federation.

Chuvash diaspora in Volga Federal District

Etymology

There is no universally accepted etymology of the word *Chuvash*, but there are two main theories that try to explain it:

- **Suvar**, according to one theory, "Chuvash" is a Shaz-Turkic adaptation of Lir-Turkic *Suvar*, an ethnonym of people that are widely considered to be the ancestors of modern Chuvashes. Compare Lir-Turkic Chuvash: *huran* to Shaz-Turkic Tatar: *qazan* (cauldron).
- **Jăvaš**, another theory suggests that the word "Chuvash" may be derived from Common Turkic *jăvaš* ("friendly", "peaceful") as opposed to *şarmăs* ("warlike").

Origin

There are rival schools of thought on the origin of Chuvash people. One is that they originated from a mixing between the Turkic Suar and Sabir tribes of Volga Bulgaria with local Mari tribes and also the Uralic[] population of the Volga Basin. Another is that the Chuvash are a remainder of the pre-Volga Bulgar population of the Volga region, merged with Scythians, Volga Bulgars and Mari.

Subgroups

Chuvash people are divided into two main groups: *Virjal* or *Turi* (Chuvash: вирьял, тури; upper) and *Anatri* (Chuvash: анатри; lower). The latter also have their own subgroups: *Anat jenci* (Chuvash: анат енчи; mid-lower) and *Hirti* (Chuvash: хирти; steppe).

History

The Turkic ancestors of the Chuvash people are believed to have come from central Siberia, where they lived in the Irtysh basin (between the Tian Shan and Altay) from at least the end of the third millennium BC.[1] In the beginning of the first century AD, the Bulgars started moving west through Zhetysu and the steppes of modern-day Kazakhstan, reaching the North Caucasus in 2nd-3rd centuries AD. There they established several states (Old Bulgaria on the Black Sea coast and the Suvar Duchy in modern-day Daghestan) and came into contact with various Iranian peoples. There is also reason to believe that they were the central element of the Hun Empire under Attila.

Old Bulgaria broke up in the second half of the 7th century after a series of successful Khazar invasions. Some of its population fled north, to the Volga-Kama region, where they established Volga Bulgaria, which eventually became extremely wealthy; its capital being the 4th largest city in the world. Shortly after that, the Suvar Duchy was forced to become a vassal state of Khazaria. About half a century later, the Suvars took part in the Khazar-Arab Wars of 732-737.

Culture

They speak the Chuvash language and have some pre-Christian traditions. In addition to the Chuvash language,[2] many Chuvash people also use the Russian and Tatar languages.

Religion

Chuvash people took great care of preserving their beliefs, their God being Tura. To prevent an onslaught of Orthodox Christianity and Islam, they separated themselves from other surrounding ethnic groups, which brought on several centuries of endogamy. Today Chuvash people are partially Orthodox Christians and belong to the Russian Orthodox Church. They retain some pre-Christian (or pagan) traditions in their cultural activities. A Pagan revival has taken place since the fall of the Soviet Union under the name Vattisen Yaly.

Gallery

Performing Chuvashes at Russian Winter Festival.

Chuvash female dancer.

Portrait of Chuvashes during a festival.

Chuvashes at the court with traditional dress.

A Chuvash woman in traditional dress.

See also

- List of Chuvashes
- Chuvash Wikipedia
- ChuvashTet

References

[1] History of Chuvash people (http://www.chuvashkrsk.org/en/) (**English**)
[2] http://www.cv-haval.org/ru/node/54 Эктор Алос-и-Фонт. Оценка языковой политики в Чувашии

Russians

The **Russians** (Russian: русские, *russkiye*) are an East Slavic ethnic group native to Russia,[1] speaking the Russian language and primarily living in Russia and neighboring countries. They are the most numerous indigenous people of the Russian Federation (more than 80% of the population, according to the census of 2010 [174]), and the most numerous people in Europe.

Terminology

There are two Russian words which are commonly translated into English as "Russians": "русские" (*russkiye*), which means "ethnic Russians" and "россияне" (*rossiyane*), which means "citizens of Russia". The first word refers to all ethnic Russians, regardless of what country they live in (Russia, Ukraine, Latvia, United States etc.) and irrespective of whether or not they hold Russian Federation citizenship, and does not include members of Russia's ethnic minorities. The second word refers to all people holding citizenship of Russia, regardless of their ethnicity, and does not include ethnic Russians living outside of Russia. English translations do not always distinguish these two words.

Origins

Further information: Rus' (people), Primary Chronicle, and Lech, Czech and Rus

The modern Russian is formed from two groups, Northern and Southern, which were made up of Krivichs, Ilmen Slavs, Radimichs, Vyatiches and Severians East Slavic tribes. Genetic studies show that modern Russians do not differ significantly from Poles or Slovenians or Ukrainians. Some ethnographers, like Zelenin, affirm that Russians are more similar to Belarusians and Ukrainians than southern Russians to northern Russians. Russians in northern European Russia share moderate genetic similarities with Uralic peoples,[] [2] who lived in modern north central European Russia and were partly assimilated by the Slavs as the Slavs migrated northeastwards. Among those peoples were Merya[3] and Muromian.[] [4]

Outside archaeological remains, little is known about the predecessors to Russians in general prior to 859 AD when the Primary Chronicle starts.[5] It is thought that by 600 AD, the Slavs had split linguistically into southern, western, and eastern branches. The eastern branch was settled between the Southern Bug and the Dnieper Rivers in what is now Ukraine; from the 1st century AD through almost the millennium, they spread peacefully northward to the Baltic region, assimilating indigents and forming the Dregovich, Radimich and Vyatich Slavic tribes on the Baltic substratum, therefore having language features such as vowel reduction. Later, both Belarusians and South Russians formed themselves on this ethnic linguistic ground.[6]

Since the 6th century, another group of Slavs moved from Pomerania to the northeast of the Baltic Sea, where they encountered the Varangians of the Rus' Khaganate and established the important regional center of Novgorod. The same Slavic ethnic population also settled the present-day Tver Oblast and the region of Beloozero. With the Uralic substratum, they formed Krivichs and Ilmen Slavs.

Anthropology of Russian

Anthropologic signs of Russian studied in detail. Russian populations are quite homogeneous in anthropological terms. Average anthropological indicators are equal to the average Western European values, or deviate from them, remaining, however, within the oscillation western groups. It is possible to note the following features that distinguish the Russian population from the Western: The proportion of light and medium shades of hair and eyes is increased, the proportion of dark hairs and eyes – reduced; Reduced growth of eyebrows and beard; Moderate width of the face; The predominance of medium and medium-high horizontal profile of the nose bridge; Smaller slope of the forehead and weaker development of brow. The Russian population is characterized by an extremely rare occurrence of epicanthus. Among more than 8,500 surveyed Russian male epicanthus found only in 12 times and its infancy only. Such an extremely rare occurrence of epicanthus observed in the German population only. According to the researches, there are two groups of Russian populations. In particular, the northern Russian on the Y-chromosome markers have a stronger resemblance to the distant Baltic States, than with more than close the Finno-Ugric peoples. On the northern Russian mtDNA resemble the gene pools of Western and Central Europe. In this case, the gene pool of the Finnish people as distant from the northern Russian. The study of autosomal markers also brings the northern Russian with other European nations, and calls into question the Finno-Ugric reservoir in the northern Russian gene pool. These data allow us to hypothesize about saving in these areas of the ancient paleoeuropean substrate that has experienced intense migration of ancient Slavic tribes. According to the research of Y-chromosome markers South central group to which the overwhelming majority of the Russian population belongs, has common cluster with Belarusians, Ukrainians and Poles. According to the researches of mtDNA markers and autosomal markers Russian similar to other populations of Central and Eastern Europe found higher unity by autosomal markers of East Slavic populations and significant differences from the neighboring Finno-Ugric, and Turkic peoples and from the North Caucasus peoples. In the Russian population there is a very low frequency of genetic characteristics of Mongoloid populations. Easteuropean frequency markers in Russian correspond to the average in Europe.

Genetics data

Russians show the characteristic R1a genes of paternal descent from a single male at 33.4% in North Russia to 49% in rest of Russia.[7] [8] [9] Such large frequencies of R1a have been found only in Eastern Europe (Sorbs, Poles and Ukrainians at about 50 to 65%),[10] [11] and in parts of India.[12]

The percentages of Y-chromosome markers vary in ethnic Russian populations, and in different studies. The top four Y-DNA haplogroups are:[13] [14]

Haplogroup R1a (Y-DNA) – 19.8% to 62.7%, with an average of **46.7%**

Haplogroup I (Y-DNA) – 0% to 26.8%, with an average of **17.6%** (All regions), and **23.5%** (Central and South Russia)

Haplogroup N (Y-DNA) – 5.4% to 53.7%, with averages of **21.6%** (All regions), and **10%** (Central and South Russia)

Haplogroup R1b (Y-DNA) – 0% to 14%, with an average of **5.8%**

Autosomally, Russians are generally similar to populations in central-eastern Europe[]

Emergence of Russian ethnicity

Ethnic origin of the first carriers of the ethnonym Rus is still being discussed. Russian ethnic group mainly was formed of the descendants of such East Slavic tribes like Krivichi, Sloveny,Vyatichi, Severiane, as well as a significant number of immigrants from the Middle Dnieper in the 12th and 13th centuries. To a lesser extent on the formation of the Russian people affected assimilation of few Finno-Ugric tribes (Meria, Mereschi, Muroma), which territories were colonized by Slavs who lived on the north-eastern territories. It should be noted that the assimilation of the Finno-Ugric tribes who lived on the Russian plain, almost had no effect on the physical type of Slavic settlers. This may be due to the proximity of the Finno-Ugric peoples of the Russian plain to the rest of the population in Eastern Europe. Notable Finno-Ugric component observed in the northern Russian, in particular, the coast-dwellers. In addition, the East Slavs, mainly Vyatichi was supposedly assimilated Balto-Slavice Golyad tribe.

Population

Russians are the most numerous ethnic group in Europe and one of the largest in the world with a population of about 140 million people worldwide. Roughly 116 million ethnic Russians live in Russia and about 16 million more live in the neighboring countries. A significant number of Russians, around 4.6 million, live elsewhere in the world, mostly in the Americas and Western Europe, but also in other places of Eastern Europe, Asia and elsewhere.

Culture

Russian culture started from that of East Slavs, who were largely polytheists, and had a specific way of life in the wooden areas of Eastern Europe. The Scandinavian Vikings, or *Varangians*, also took part in the forming of Russian identity and state in the early Kievan Rus' period of the late 1st millennium AD. Rus' had accepted the Christianity from the East Roman Empire in 988, and this largely defined the Russian culture of next millennium as the synthesis of Slavic and Byzantine cultures.[] After the fall of Constantinople in 1453, Russia remained the largest Orthodox nation in the world and claimed succession to the Byzantine legacy in the form of the Third Rome idea. At different points of its history, the country also was strongly influenced by the European Culture, and since Peter the Great reforms Russian culture largely developed in the context of the Western culture. For most of the 20th century, the Marxist ideology shaped the culture of the Soviet Union, where Russia, or Russian SFSR, was the largest and leading part.

Saint Basil's Cathedral on the Red Square, Moscow

Russian culture is extremely various and unique in many aspects. It has a rich history and can boast a long tradition of excellence in every aspect of arts,[] especially when it comes to literature[15] and philosophy, classical music[16] [17] and ballet,[18] architecture and painting, cinema[] and animation, which all had considerable influence on the world culture.

Russian literature is known for such notable writers as Aleksandr Pushkin, Leo Tolstoy, Fyodor Dostoyevsky, Anton Chekhov, Vladimir Mayakovsky, Boris Pasternak, Anna Akhmatova, Joseph Brodsky, Maxim Gorky, Vladimir Nabokov, Mikhail Sholokhov, Mikhail Bulgakov, Andrei Platonov, Aleksandr Solzhenitsyn, and Varlam Shalamov. Russians also gave the classical music world some very famous composers, including Piotr Ilyich Tchaikovsky and his contemporaries, the Mighty Handful, including Modest Mussorgsky and Nikolai Rimsky-Korsakov. In the 20th-century Russian music was credited with such influential composers as Dmitri Shostakovich, Sergei Prokofiev,

Sergei Rachmaninoff, Igor Stravinski, Georgy Sviridov, and Alfred Schnittke. Many more famous Russian people are associated with different aspects of culture.

Russian philosophy

According to Lossky, the characteristic features of Russian philosophy are: Space Art, sophiology (the doctrine of Sophia), collegiality, metaphysical, religious, intuitionism, positivism, ontologism. The subjects of philosophical search for Russian thinkers were: The problem of man; Space Art (perception of space as a single whole organism); Issues of morality and ethics; The problem of choosing the historical path of Russia – between East and West (specific problem of Russian philosophy, which led to a confrontation Russian intelligentsia in the middle of the 19th century: see Slavophilism, Westernism); The problem of power; The problem of the state; The problem of social justice; The problem of an ideal society; The problem of the future.

Language

Russian (русский язык , transliteration: *Russkiy yazyk*, Russian pronunciation: [ˈruskʲɪj jɪˈzɨk]) is the most geographically widespread language of Eurasia and the most widely spoken of the Slavic languages. Russian belongs to the family of Indo-European languages and is one of three (or, according to some authorities, four) living members of the East Slavic languages, the others being Belarusian and Ukrainian.

Examples of Old East Slavonic are attested from the 10th century onwards, and while Russian preserves much of East Slavonic grammar and a Common Slavonic word base, modern Russian exhibits a large stock of borrowed international vocabulary for politics, science, and technology. Due to the status of the Soviet Union as a super power, Russian had great political importance in the 20th century, and is one of the official languages of the United Nations.

Russian has palatal secondary articulation of consonants, the so-called *soft* and *hard* sounds. This distinction is found in almost all consonant phonemes and is one of the most distinguishing features of the language. Another important aspect is the reduction of unstressed vowels, not entirely unlike a similar process present in most forms of English. Stress in Russian is generally quite unpredictable and can be placed on almost any syllable, one of the most difficult aspects for foreign language learners.

A group of Russian children, 1909. Sergei Mikhailovich Prokudin-Gorskii.

Russian language – one of the six official languages of the UN. According to data published in the journal «Language Monthly» (№ 3, 1997), approximately 300 million people around the world at the time owned the Russian language (which puts this language in 5th place on prevalence in the World). 160 million considered Russian as native language (7th in the world). The total number of Russian speakers in the world on the 1999 assessment – about 167 million, about 110 million people speak Russian as a second language. In a sociological study of Gallup (Gallup, Inc), dedicated to the Russian language in the post-Soviet states, 92% of the population in Belarus, 83% in Ukraine, 68% in Kazakhstan and 38% in Kyrgyzstan, Russian language chosen to complete the questionnaire for the survey . Institute designated this section of the study as «Russian as the Mother Tongue». In the U.S. state of New York in 2009, an amendment to the electoral law, according to which in all cities in the state, home to over a million people, all related to the election process documents should be translated into Russian. Russian language has become one of the eight foreign languages in New York, which must be printed on all official materials of the campaign. Previously been included in the list of Spanish, Korean, Filipino, Creole languages and three dialects of Chinese.

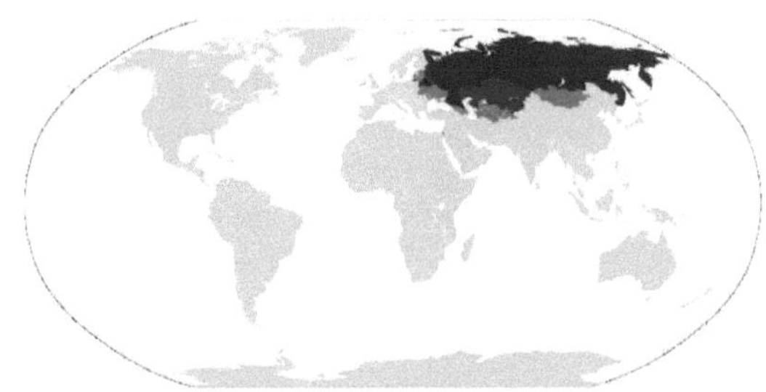

Russian language has official status. Russian language has official status, but lower than the state Russian language has no official status, but is known with much of the population.

Prior to 1991, Russian was the language of international communication of the USSR, the de facto fulfilling the functions of the state language. Continues to be used in countries formerly part of the Soviet Union, as a mother for a significant portion of the population and as a language of international communication. In places of compact residence of immigrants from the former USSR countries (Israel, Germany, Canada, the United States, Australia, etc.) are available Russian-language periodicals, work stations and television channels, are opened Russian-language schools, where actively taught Russian. In the countries of Eastern Europe before the end of the 1980s Russian language was the main foreign language taught in schools. All astronauts working in International Space Station, should speak spoken Russian language.

Religion

Around 63% of the Russia's population identify themselves with Orthodox Christianity,[19] most of whom belong to the Russian Orthodox Church, which played a vital role in the development of Russian national identity. In other countries Russian faithful usually belong to the local Orthodox congregations which either have a direct connection (like the Ukrainian Orthodox Church, autonomous from the Moscow Patriarchate) or historical origin (like the Orthodox Church in America or a Russian Orthodox Church Outside of Russia) with the Russian Orthodox Church.

Non-religious Russians may associate themselves with the Orthodox faith for cultural reasons. Some Russian people are Old Believers: a relatively small schismatic group of the Russian Orthodoxy that rejected the liturgical reforms introduced in the 17th century. Other schisms from Orthodoxy include Doukhobors which in the 18th century rejected secular government, the Russian Orthodox priests, icons, all church ritual, the Bible as the supreme source of divine revelation and the divinity of Jesus, and later emigrated into Canada. An even earlier sect were Molokans which formed in 1550 and rejected Czar's divine right to rule, icons, the Trinity as outlined by the Nicene Creed, Orthodox fasts, military service, and practices including water baptism.

Other world religions have negligible representation among ethnic Russians. The largest of these groups are Islam with over 100,000 ethnic Russian followers[20] , and Baptists with over 85,000 Russian adherents.[21] Others are mostly Pentecostals, Evangelicals, Seventh-day Adventists, Lutherans and Jehovah's Witnesses.

Since the fall of the Soviet Union various new religious movements have sprung up and gathered a following among ethnic Russians. The most prominent of these are Rodnovery, the revival of the Slavic native religion also common to other Slavic nations,[22] Another movement, very small in comparison to other new religions, is Vissarionism, a syncretic group with an Orthodox Christian background.

Russians outside of Russia

Ethnic Russians historically migrated throughout the area of former Russian Empire and Soviet Union, sometimes encouraged to re-settle in borderlands by Tsarist and later Soviet government.[23] On some occasions ethnic Russian communities, such as Lipovans who settled in the Danube delta or Doukhobors in Canada, emigrated as religious dissidents fleeing the central authority.

Sainte-Geneviève-des-Bois Russian Cemetery, the resting place of many eminent Russian émigrés after 1917

After the Russian Revolution and Russian Civil War starting in 1917, many Russians were forced to leave their homeland fleeing the Bolshevik regime, and millions became refugees. Many white émigrés were participants in the White movement, although the term is broadly applied to anyone who may have left the country due to the change in regime.

Today the largest ethnic Russian diasporas outside of Russia live in former Soviet states such as Ukraine (about 8 million), Kazakhstan (about 3.8 million), Belarus (about 785,000), Latvia (about 556,000) with the most Russian settlement out of the Baltic States which includes Lithuania and Estonia, Uzbekistan (about 650,000) and Kyrgyzstan (about 419,000).

Russian Orthodox Church in Shanghai around 1948

Over a million Russian Jews emigrated to Israel during and after the Refusenik movements; some brought ethnic Russian relatives along with them. Out of more than one million Russian-speaking immigrants in Israel,[24] about 300,000 are considered not Jewish according to the rabbinical commandments (but not all of them are ethnic Russians).[25] There are also small Russian communities in the Balkans, Eastern and Central European nations such as Germany and Poland, as well Russians settled in China, Japan, South Korea, Latin America (i.e. Mexico, Brazil and Argentina) and Australia. These communities may identify themselves either as Russians or citizens of these countries, or both, to varying degrees.

People who had arrived in Latvia and Estonia during the Soviet era, including their descendants born in these countries, mostly Russians, became stateless after the dissolution of the Soviet Union and were provided only with an option to acquire naturalised citizenship. The language issue is still contentious, particularly in Latvia, where ethnic Russians have protested against plans to liquidate education in minority languages, including Russian. Since 1992, Estonia has naturalized some 137,000 residents of undefined citizenship, mainly ethnic Russians. 136,000, or 10 percent of the total population, remain without citizenship.

Both the European Union and the Council of Europe, as well as the Russian government, expressed their concern during the 1990s about minority rights in several countries, most notably Latvia and Estonia. In Moldova, the Transnistria region (where 30.4% of population is Russian) broke away from government control amid fears the country would soon reunite with Romania. In June 2006, Russian President Vladimir Putin announced the plan to introduce a national policy aiming at encouraging ethnic Russians to immigrate to Russia.[26]

Ethnic Russians in former Soviet Union states

Significant numbers of Russians emigrated to Canada, Australia and the United States. Brighton Beach, Brooklyn and South Beach, Staten Island in New York City is an example of a large community of recent Russian and Jewish Russian immigrants. Other examples are Sunny Isles Beach, a northern suburb of Miami, and "Little Moscow" in Hollywood of the Los Angeles area.

At the same time, many ethnic Russians from former Soviet territories have emigrated to Russia itself since the 1990s. Many of them became refugees from a number of states of Central Asia and Caucasus (as well as from the separatist Chechen Republic), forced to flee during political unrest and hostilities towards Russians.

After the Russian Revolution in 1917, many Russians who were identified with the White army moved to China — most of them settling in Harbin and Shanghai. By the 1930s Harbin had 100,000 Russians. Many of these Russians had to move back to the Soviet Union after World War II. Today, a large group of people in northern China can still speak Russian as a second language.

Russians (*eluosizu*) are one of the 56 ethnic groups officially recognized by the People's Republic of China (as *the Russ*); there are approximately 15,600 Russian Chinese living mostly in northern Xinjiang, and also in Inner Mongolia and Heilongjiang.

Notable achievements

Various Russians have greatly contributed to the world of music, sports, science, technology and arts. Notable Russian scientists include Dmitri Mendeleev, Nikolay Bogolyubov, Andrei Kolmogorov, Ivan Pavlov, Nikolai Semyonov, Dmitri Ivanenko, Nikolai Lobachevsky, Alexander Lodygin, Alexander Popov (one of inventors of radio), Nikolai Zhukovsky, Alexander Prokhorov and Nikolay Basov (co-inventors of laser), Georgiy Gamov, Vladimir Zworykin, Lev Pontryagin, Sergei Sobolev, Pavel Yablochkov, Aleksandr Butlerov, Andrei Sakharov, Dmitry Ivanovsky, Sergey Korolyov and Mstislav Keldysh (creators of the Soviet space program), Aleksandr Lyapunov, Mikhail Dolivo-Dobrovolsky, Andrei Tupolev, Yuri Denisyuk (the first practicable method of holography), Mikhail Lomonosov, Vladimir Vernadsky, Pyotr Kapitsa, Igor Sikorsky, Ludvig Faddeev, Zhores Alferov, Konstantin Novoselov, Fyodor Shcherbatskoy, Nikolai Trubetzkoy etc.

Yuri Gagarin, first human in space (1961)

The first man in space, Yuri Gagarin, was Russian, and the first artificial satellite to be put into outer space, Sputnik 1, was launched by the Soviet Union and was developed mainly by Sergey Korolyov who had a Russian father (his mother was Ukrainian).

Russian Literature representatives like Fyodor Dostoyevsky, Lev Tolstoy, Ivan Turgenev, Anton Chekhov, Alexander Pushkin, and many more, reached a high status in world literature. In the field of the novel, Tolstoy and Dostoyevsky, in particular, were important figures and have remained internationally renowned. Some scholars have described one or the other as the greatest novelist ever.[27]

Russian composers who reached a high status in the world of music include Igor Stravinsky, Pyotr Ilyich Tchaikovsky, Dmitri Shostakovich, Nikolai Rimsky-Korsakov, Sergei Prokofiev, and Sergei Rachmaninoff.

Russian people played a crucial role in the victory over Nazi Germany in World War II. Russia's casualties in this war were the highest of all nations, and numbered more than 20 million dead (Russians composed 80% of the 26.6 million people lost by the USSR), which is about half of all World War II casualties and the vast majority of Allied casualties.[28] According to the British historian Richard Overy, the Eastern Front included more combat than all the other European fronts combined. The Wehrmacht suffered 80% to 93% of all of its total World War II combat casualties on the Eastern Front.

See also

- List of Russians
- List of Russian artists
- List of Russian inventors
- Timeline of Russian inventions and technology records
- Russian culture
- Russian diaspora
- European ethnic groups

References

[1] "Slav." (http://www.britannica.com/EBchecked/topic/548156/Slav) Encyclopædia Britannica Online. 30 July 2011.
[2] Two Sources of the Russian Patrilineal Heritage in Their Eurasian Context (http://www.pubmedcentral.nih.gov/articlerender.fcgi?artid=2253976) by Oleg Balanovsky, Siiri Rootsi, Andrey Pshenichnov, Toomas Kivisild et al.
[3] *Aleksey Uvarov, "Étude sur les peuples primitifs de la Russie. Les mériens" (1875)
[4] (http://www.emc.komi.com/01/12/115.htm)
[5] The Primary Chronicle is a history of the Ancient Rus' from around 850 to 1110 originally compiled in Kiev about 1113)
[6] Pivtorak. Formation and dialectal differenciaton of the Old Rus language. 1988
[7] Rosser et al. (2000)
[8] Tambets et al. (2004).
[9] Oleg Balanovsky, Siiri Rootsi, Andrey Pshenichnov et al., "Two Sources of the Russian Patrilineal Heritage in Their Eurasian Context," American Journal of Human Genetics 82, 236–250, January 2008.
[10] Balanovsky et al. (2008), "Two Sources of the Russian Patrilineal Heritage in Their Eurasian Context"
[11] Semino et al. (2000), "The Genetic Legacy of Paleolithic Homo sapiens sapiens in Extant Europeans: A Y Chromosome Perspective"
[12] F. Luca, F. Di Giacomo, T. Benincasa et al., "Y-Chromosomal Variation in the Czech Republic," American Journal of Physical Anthropology 132:132–139 (2007).
[13] http://ruthen-info.fatal.ru/files/mtDNA_Y_Chromosomes_in_Russian_Populations.pdf
[20] 1.^ http://sreda.org/arena/arena-v-pdf
[22] Victor Shnirelman. *"Christians! Go home": A Revival of Neo-Paganism between the Baltic Sea and Transcaucasia* (http://www.wlu.ca/documents/6483/Christians_Go_home.pdf). Journal of Contemporary Religion, Vol. 17, No. 2, 2002.
[23] Russians left behind in Central Asia (http://news.bbc.co.uk/2/hi/asia-pacific/4420922.stm). BBC News. November 23, 2005.
[24] Study: Soviet immigrants outperform Israeli students (http://www.haaretz.com/hasen/spages/952558.html). Haaretz.com. 10/02/2008.
[25] Q&A Lily Galili on 'The Russians in Israel' (http://www.haaretz.com/hasen/pages/QA.jhtml?qaNo=59). Haaretz.com
[27] "Russian literature." Encyclopædia Britannica. 2007. Encyclopædia Britannica Online. 16 July 2007 <http://www.britannica.com/eb/article-29157>.
[28] Leaders mourn Soviet wartime dead (http://news.bbc.co.uk/1/hi/world/europe/4530565.stm), BBC News

External links

- (**English**) Prominent Russians: faces of Russia (http://russiapedia.rt.com/prominent-russians/) – Russia Today
- (**Russian**) 4.1. Population by nationality (http://www.perepis2002.ru/ct/html/TOM_14_24.htm)
- (**Russian**) "People and Cultures: Russians" (http://www.booksite.ru/fulltext/rus/sian/index.htm) book published by Russian Academy of Sciences
- (**English**) China Internet Information Center – The Russian Ethnic Group (http://service.china.org.cn/link/wcm/Show_Text?info_id=139526&p_qry=russians/)
- Russkiy Mir Foundation (http://www.russkiymir.ru/russkiymir/en/)
- Russian folk costumes (female) (http://www.ruplace.ru/fotogalereya/russkiy-narodnyy-kostyum-v-rabotah-dmitriya-davydova/belgorodskaya-oblastj.html#ponyimg)
- Pre-Revolutionary photos of women in Russian folk costumes (http://folkportal.3dn.ru/forum/17-186-1)

Time_in_Russia

There are nine **time zones in Russia**, which currently observe times ranging from UTC+03:00 to UTC+12:00. UTC+05:00 is not used.

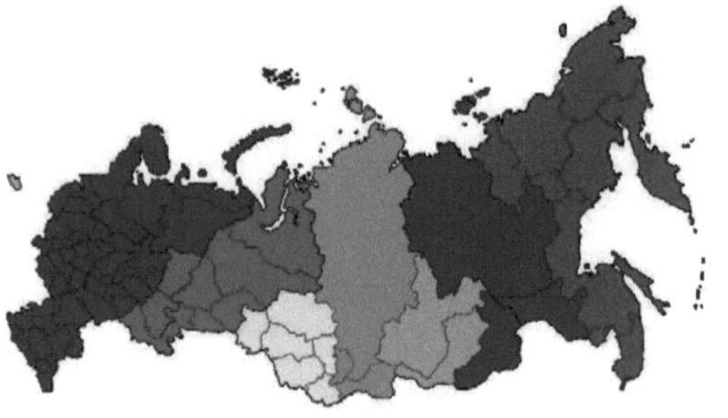

Time in Russia

List of zones

Since March 2011, the time zones are as follows:

Time Zone Name	Time of day and abbreviation(s)	UTC offset	Area covered
Kaliningrad Time	()	UTC+03	Kaliningrad Oblast (located in Europe)
Moscow Time	()	UTC+04	Most of European Russia
Yekaterinburg Time	()	UTC+06	Bashkortostan, Chelyabinsk Oblast, Khanty–Mansi Autonomous Okrug, Kurgan Oblast, Orenburg Oblast, Perm Krai, Sverdlovsk Oblast (partly in Europe), Tyumen Oblast, and Yamalia
Omsk Time	()	UTC+07	Altai Krai, Altai Republic, Novosibirsk Oblast, Kemerovo Oblast, Omsk Oblast and Tomsk Oblast
Krasnoyarsk Time	()	UTC+08	Khakassia, Krasnoyarsk Krai and Tuva
Irkutsk Time	()	UTC+09	Irkutsk Oblast and Buryatia
Yakutsk Time	()	UTC+10	Amur Oblast, western Sakha Republic and Zabaykalsky Krai
Vladivostok Time	()	UTC+11	Jewish Autonomous Oblast, Khabarovsk Krai, Primorsky Krai, central Sakha Republic and Sakhalin Oblast (except its easternmost district in the Kuril Islands)
Magadan Time	()	UTC+12	Magadan Oblast, eastern Sakha Republic, Severo-Kurilsky District of the Sakhalin Oblast, Chukotka and Kamchatka Krai

Daylight saving time

On 8 February 2011, Russian President Dmitry Medvedev issued a decree that Russia would observe year-round DST. Under the decree, all clocks in Russia were advanced one hour on 27 March 2011 as usual, but are not to change back the following October, effectively making Moscow Time UTC+4 permanently.[]

History of zone boundaries

In the Russian Empire, most of the nation observed solar time. During the late 19th century, Moscow Mean Time was introduced, originally at UTC+02:30. However, when the Soviet Union was created, Moscow Time became UTC+02 and the various other time zones were introduced throughout Russia and the rest of the Soviet Union. On 21 June 1930, the Soviet Union advanced all clocks by one hour, effectively making the nation run on daylight saving time all year. Russia and most republics in the Soviet Union turned their clocks back one hour on 31 March 1991, but Russia reversed this the following year.

In 1992, the Government of Russia issued a resolution establishing the borders of the eleven time zones dividing the country's territory.[1] During the following years, various clauses of the resolution were superseded by other laws, re-assigning various federal subjects to different time zones.

The only federal subjects to be in more than one time zone are the Sakha Republic, which is separated into areas which belong to the Yakutsk, Vladivostok and Magadan time zones, and Sakhalin Oblast, which is in the Vladivostok (Sakhalin Island, and the two westernmost districts in the Kuril Islands) and Magadan time zones (Severo-Kurilsky District in the Kuril Islands).

In November 2009, President of Russia Dmitry Medvedev proposed reducing the number of time zones spanning the country,[] as well as the abolition of daylight saving time.

1993 zone boundary changes

On 23 May 1993 00:00:00, Novosibirsk Oblast changed its time zone from MSK+4 to MSK+3.[2]

The change occurred during DST effectively changing the offset from UTC+8 to UTC+7, the offset without DST was therefore changed from UTC+7 to UTC+6.

2002 zone boundary changes

In 2002, Tomsk Oblast changed its time zone from MSK+4 to MSK+3.[3] [4] [5]

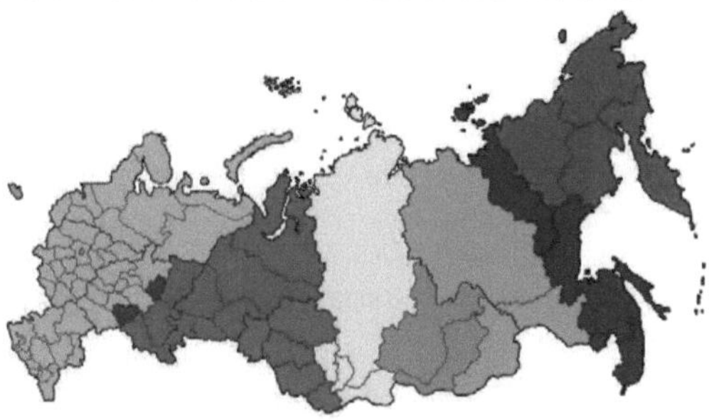

11 time zones in Russia from 2002 to 2010

2010 zone boundary changes

On 28 March 2010, the following changes were introduced, which, in particular, led to abolition of two of the eleven time zones.

- The Udmurt Republic and Samara Oblast started using Moscow Time, thus eliminating Samara Time (MSK+1 or UTC+4 without DST).[6] [7]
- Kemerovo Oblast started using Omsk Time.[8]
- Chukotka Autonomous Okrug and Kamchatka Krai started using Magadan Time, thus eliminating Kamchatka Time (MSK+9 or UTC+12 without DST).[9]

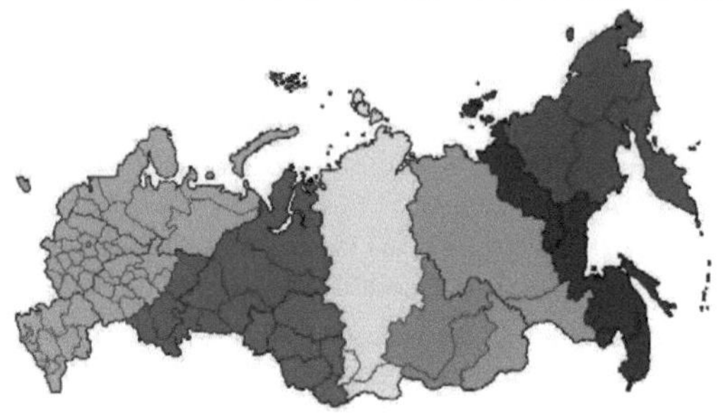

April 2010: 9 zones

Although the Russian government wants to reduce the number of time zones even further, there have been protests in far-eastern Russia on the recent changes, including protests and a 20,000-strong petition in support of Kamchatka returning to UTC+12.[10]

2011 zone boundary changes

The decree No. 725 [11] (31 August 2011) defines Moscow Time and lists the zones, numbered 1 to 9 and in turn defined relative to Moscow Time. The offsets from UTC are altered, the reference zone, Moscow Time Zone, now uses UTC+04:00. This is similar to the 1930 USSR reform with decree time coming into operation.

As a result some districts of the Sakha Republic switched from Vladivostok Time to Yakutsk Time (Zone 7):

- Tomponsky District
- Ust-Maysky District

September 2011: 9 zones, "permanent DST".

Some districts switched from Magadan Time to Vladivostok Time:

- Oymyakonsky District (Sakha Republic)
- Kurilsky District and Yuzhno-Kurilsky District (Sakhalin Oblast)

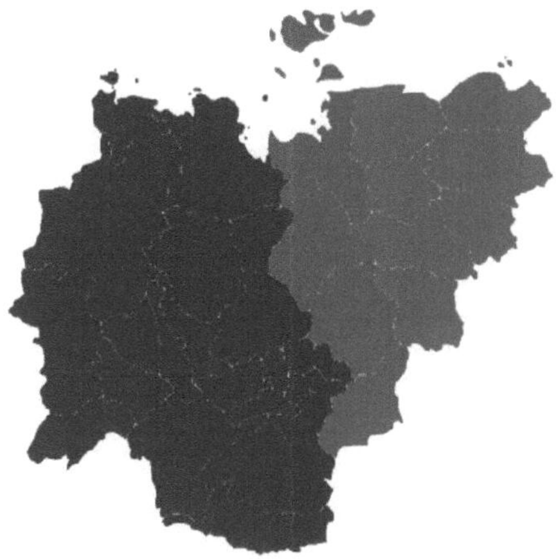

1 April 2010 (with DST)

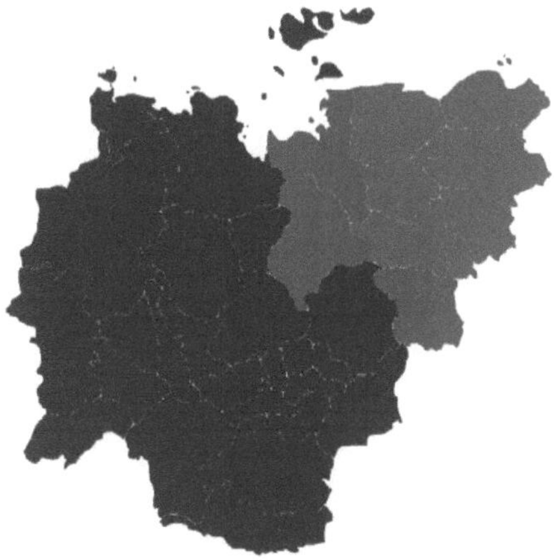

1 September 2011 ("permanent DST")

Blue Yakutsk Time (MSK+6), pink Vladivostok Time (MSK+7), red Magadan Time (MSK+8).

Railway time

All timetables on Russian Railways (except Sakhalin railways) follow Moscow Time.[12] Airports, however, follow local time.[13]

tz database

For Russia, the tz database contains several zones in the file zone.tab.

http://efele.net/maps/tz/russia - data from 2009

List of zones

The list below shows the 16 zones for Russia as defined in the file zone.tab of the database. The database aims to identify regions that had the same time offset rules since 1970.

Two federal subjects are contained in more than one tz zone. The Sakha Republic is divided into three: west, central, east. Sakhalin Oblast is divided into two: Sakhalin Island with Kurilsky and Yuzhno-Kurilsky districts in the Kuril Islands, and Severo-Kurilsky District in the Kuril Islands.

Two zones, namely Asia/Omsk and Asia/Novosibirsk, each cover area that did not observe the same rule set since 1970, all now using Omsk Time.

On the last Sunday in October 2011, daylight-saving time ended in tzdata, but all zones moved forward one hour. In other words, the clocks did not change, but the names of the time zones reverted permanently to their standard time variants and there will be no more daylight-saving time going forward.

If available, the change column lists the offset changes that caused a creation of a new zone in the tz database. "Initial zone" means that in 1970 there was already a difference in time offset from the offsets in any other zone.

C.c.	Coordinates	tzid	Comments	UTC offset (without DST, permanent since 2011)	Covered area	Split from	Changes
RU	+5443+02030	Europe/Kaliningrad	Moscow-01 - Kaliningrad	+03:00	Kaliningrad Oblast	Initial zone	
RU	+5545+03735	Europe/Moscow	Moscow+00 - west Russia	+04:00	Most of European Russia. Complete list given here.	Initial zone	
RU	+4844+04425	Europe/Volgograd	Moscow+00 - Caspian Sea	+04:00	Kirov Oblast, Saratov Oblast, Volgograd Oblast, and Astrakhan Oblast	Europe/Samara	1992-03-29 Zone creation, causing change from UTC+04 to UTC+03
RU	+5312+05009	Europe/Samara	Moscow+00 - Samara, Udmurtia	+04:00	Samara Oblast and Udmurtia	Initial zone	2010-03-28 Change from UTC+04 to UTC+03
RU	+5651+06036	Asia/Yekaterinburg	Moscow+02 - Urals	+06:00	Bashkortostan, Chelyabinsk Oblast, Khanty–Mansi Autonomous Okrug, Kurgan Oblast, Orenburg Oblast, Perm Krai, Sverdlovsk Oblast, Tyumen Oblast, and Yamalia	Initial zone	
RU	+5500+07324	Asia/Omsk	Moscow+03 - west Siberia	+07:00	Altai Krai, Altai Republic, and Omsk Oblast		• 1995-05-28 Zone creation, causing change from UTC+07 to UTC+06 • 1992-01-19 Change from UTC+05 to UTC+06
RU	+5502+08255	Asia/Novosibirsk	Moscow+03 - Novosibirsk	+07:00	Novosibirsk Oblast and Tomsk Oblast.		• 1993-05-23 Zone creation, causing change from UTC+07 to UTC+06 • 2002-05-01 Change from UTC+07 to UTC+06[3] [4] [5]
RU	+5345+08707	Asia/Novokuznetsk	Moscow+03 - Novokuznetsk	+07:00	Kemerovo Oblast	Asia/Novosibirsk	2010-03-28 Zone creation, causing change from Krasnoyarsk Time to Novosibirsk Time[14]

RU	+5601+09250	Asia/Krasnoyarsk	Moscow+04 - Yenisei River	+08:00	Khakassia, Krasnoyarsk Krai, and Tuva Republic		
RU	+5216+10420	Asia/Irkutsk	Moscow+05 - Lake Baikal	+09:00	Irkutsk Oblast and Buryatia		
RU	+6200+12940	Asia/Yakutsk	Moscow+06 - Lena River	+10:00	Amur Oblast, Zabaykalsky Krai, and western Sakha Republic		
RU	+4310+13156	Asia/Vladivostok	Moscow+07 - Amur River	+11:00	Jewish Autonomous Oblast, Khabarovsk Krai, Primorsky Krai, and central Sakha Republic	Initial zone	
RU	+4658+14242	Asia/Sakhalin	Moscow+07 - Sakhalin Island	+11:00	Sakhalin Island, and western Kuril Islands	Asia/Magadan	1997-03-30 Zone creation, causing change from UTC+11 to UTC+10
RU	+5934+15048	Asia/Magadan	Moscow+08 - Magadan	+12:00	Magadan Oblast, eastern Kuril Islands, and eastern Sakha Republic		
RU	+5301+15839	Asia/Kamchatka	Moscow+08 - Kamchatka	+12:00	Kamchatka Krai	Initial zone	2010-03-28 Change from UTC+12 to UTC+11
RU	+6445+17729	Asia/Anadyr	Moscow+08 - Bering Sea	+12:00	Chukotka Autonomous Okrug	Initial zone	• 1982-04-01 Changed from UTC+13 to UTC+12 • 2010-03-28 Changed from UTC+12 to UTC+11

Deleted zones

Asia/Ulan Ude was a time zone identifier from the zone file of the tz database. The reference point was Ulan-Ude. It was added in tz version 2011e.[15]

Federal subjects with multiple offsets at the same time

Per a 2011 law[16] the territories of Sakhalin Oblast and Sakha Republic each observe more than one offset.

Sakha Republic:

- UTC+10 MSK+6 Asia/Yakutsk
- UTC+11 MSK+7 Asia/Vladivostok
- UTC+12 MSK+8 Asia/Magadan

Sakhalin Oblast:

- UTC+11 MSK+7 All except Severo-Kurilsky District in the Kuril Islands: Asia/Sakhalin
- UTC+12 MSK+8 Severo-Kurilsky District in the Kuril Islands: Asia/Magadan

See also

- Decree time

References

[1] Правительство Российской Федерации. Постановление №23 от 8 января 1992 г. «О порядке исчисления времени на территории Российской Федерации». (Government of the Russian Federation. Resolution #23 of 8 January 1992 *On the Procedures of Keeping Time on the Territory of the Russian Federation.*).

[2] http://www.timeanddate.com/worldclock/clockchange.html?n=375&year=1993

[3] http://article.gmane.org/gmane.comp.time.tz/3626/match=asia+tomsk

[4] http://newsru.com/russia/01may2002/clock.html

[5] http://www.timegenie.com/state.time/rutom

[6] Правительство Российской Федерации. Постановление №166 от 17 марта 2010 г. « О применении на территории Удмуртской Республики времени второго часового пояса (http://www.consultant.ru/online/base/?req=doc;base=LAW;n=98575)». Опубликован: "Российская Газета", №58, 22 марта 2010 г. (Government of the Russian Federation. Resolution #166 of 17 March 2010 *On Using the Time of the Second Time Zone on the Territory of the Udmurt Republic* (http://translate.google.com/translate?hl=en&sl=ru&tl=en&u=http://www.consultant.ru/online/base/?req=doc;base=LAW;n=98575).).

[7] Правительство Российской Федерации. Постановление №170 от 19 марта 2010 г. « О применении на территории Самарской области времени второго часового пояса (http://www.consultant.ru/online/base/?req=doc;base=LAW;n=98576)». Опубликован: "Российская Газета", №58, 22 марта 2010 г. (Government of the Russian Federation. Resolution #170 of 19 March 2010 *On Using the Time of the Second Time Zone on the Territory of Samara Oblast* (http://translate.google.com/translate?hl=en&sl=ru&tl=en&u=http://www.consultant.ru/online/base/?req=doc;base=LAW;n=98576).).

[8] Правительство Российской Федерации. Постановление №740 от 14 сентября 2009 г. «О применении на территории Кемеровской области времени пятого часового пояса». (Government of the Russian Federation. Resolution #740 of 14 September 2009 *On Using the Time of the Fifth Time Zone on the Territory of Kemerovo Oblast.*).

[9] Правительство Российской Федерации. Постановление №171 от 19 марта 2010 г. « О применении на территории Камчатского края и Чукотского автономного округа времени десятого часового пояса (http://www.consultant.ru/online/base/?req=doc;base=LAW;n=98577)». Опубликован: "Российская Газета", №58, 22 марта 2010 г. (Government of the Russian Federation. Resolution #171 of 19 March 2010 *On Using the Time of the Tenth Time Zone on the Territory of Kamchatka Krai and Chukotka Autonomous Okrug* (http://translate.google.com/translate?hl=en&sl=ru&tl=en&u=http://www.consultant.ru/online/base/?req=doc;base=LAW;n=98577).).

[11] Постановление Правительства Российской Федерации № 725 от 31 августа 2011 г. «О составе территорий, образующих каждую часовую зону, и порядке исчисления времени в часовых зонах, а также о признании утратившими силу отдельных постановлений Правительства Российской Федерации». (http://www.rg.ru/2011/09/06/chas-zona-dok.html)

[12] Russian Railways - Time tables (http://eng.rzd.ru/isvp/public/rzdeng/express?STRUCTURE_ID=46)

[13] for example http://www.iktport.ru/

[14] http://article.gmane.org/gmane.comp.time.tz/2955/match=novokuznetsk

[15] http://article.gmane.org/gmane.comp.time.tz/3683/match=asia+ulan_ude

External links

- Map of time zones in Russia (http://www.worldtimezone.com/time-russia24.php)
- Time Zones of Russia (http://www.statoids.com/tru.html)

Article Sources and Contributors

Yelovsky_District *Source*: http://en.wikipedia.org/w/index.php?title=Yelovsky_District *Contributors*: Brainwashinguser, Ezhiki, Rich Farmbrough, Starzynka

Perm_Krai *Source*: http://en.wikipedia.org/w/index.php?title=Perm_Krai *Contributors*: Abhijitsathe, Anandks007, Backspace, Barend, Brainwashinguser, Bubba Zanetty, CarolSpears, Chanheigeorge, Citterio, Conscious, Courcelles, Crusoe8181, Ettrig, Ezhiki, Fasmike3, Flrn, Garant^^, Gobi go, Good Olfactory, Ground Zero, Grue, HarDNox, Hunnjazal, Imc, Iwillremembermypassthistime, JamesAM, Kintetsubuffalo, Kiril Simeonovski, KirkEN, Kuban kazak, Lady Pablo, LodeRunner, Magioladitis, Maximaximax, Nightstallion, Nikai, Paul A, Rdfr, Rich Farmbrough, Rjwilmsi, Russavia, SQGibbon, Sardanaphalus, Schekinov Alexey Victorovich, Seb az86556, Sharveet, Sombrero, Superzohar, Therexbanner, Tibetan Prayer, Urmas, Vuvar1, WereSpielChequers, Wheelerwiki, Yannzgob, Пакко, 40 anonymous edits

Russia *Source*: http://en.wikipedia.org/w/index.php?title=Russia *Contributors*: 'Ndrangheta, *drew, -- April, .:Ajvol:., .mdk., 08OceanBeach SD, 12 Noon, 1234r00t, 1297, 172, 1brettsnyder, 1exec1, 212.192.249.xxx, 24630, 25 Cents FC, 2812, 31ff3y, 32167, 334a, 6 million jews, 66 red hat 66, 747fzx, A Werewolf, A bit iffy, A man without a country, A-giau, A.Savin, A.h. king, A1916, A3r0, AOEU, ARC Gritt, ARUenergy, ASDFGH, Aa75253, Aaker, Aardark, Aaron Einstein, Aaron danielg, AaronPaige, Aaroncrick, Abbadonnergal, Abbeyroad55343, Abce2, Abdominator, Aborig, Abune, Academic Challenger, Accurizer, Ace of Raves, Acer, Acerperi, Acers12, Acio83, Acs4b, AdRock, Adam Carr, Adam78, AdamFouracre, Adambiswanger1, Adashiel, Adeist, Adieujuste, AdjustShift, AdrianTM, Aec is away, Aecis, Aero145, Aesopos, Aeusoes1, Afinogenoff, Afrasclient, Ageekgal, Agent 78787, Agent452, Agomulka, Ahh im going bald, Ahoerstemeier, Ahuskay, Aibur, Aillema, Airplaneman, Aivazovsky, Ajmint, Ajraddatz, Akanemoto, Akbg, Akhristov, Akshaysrinivasan, Al3xil, Alaexis, Alam82, Alaney2k, Alangstone, Alasdairking, Albanaco, Aleenf1, Aleksei, Alethiophile, Alex '05, Alex Bakharev, Alex earlier account, Alex756, AlexNebraska, AlexPU, Alexander Domanda, Alexander Gerashchenko, AlexanderKaras, Alexikoua, AlexiusHoratius, Alexnevzorov, Ali, Aliyevramin, Allan Letinov, AllenHansen, Alloy, Allysia, Alnrr, Alphachimp, Alphasinus, Altau, Altenmann, Alton, Alynna Kasmira, Amberrock, Ameliorate!, Amenzix, Amire80, Amrad, Amylyn, Anandks007, Anddrex, Andonic, Andonio, Andre Engels, Andres, AndrewDressel, AndrewHowse, Andrijko Z., Andris, AndriyK, Android79, Androidsdungeon, Andy Marchbanks, Andy120290, Andyjsmith, AndytheTweedale, Ang15, Angelo De La Paz, Anger22, Angr, Angusmclellan, Anonymous from the 21st century, Anonymous from the 21th century, Anonymous44, Answerthis, Antandrus, Anthonyd3ca, Antipode, Anton Fischer, Anton Khorev, Antonio Basto, Anubis-SG, Anvyder, Aogouguo, Aphaia, Apoivre, Appleboy, Aprolix, Aramgutang, Arbor to SJ, Arcenciel, Archer3, Archiver of Records, Arem04667, Ares1991, Aria613, Arichnad, Arigato1, Armanos, Armydepot, Arne List, Aromatkzn, Art LaPella, Arthur889, Article editor, Arturo313, Arun11, Asdfking1, Ash sul, Asidemes, Assassin3577, Assyrial, Astronaut, Astropithicus, Astrotrain, Atilano, AtilimGunesBaydin, Atitarev, Atlantas, Atoric, Aude, AugustinMa, Auntof6, Ausir, Avala, Avaya1, AvicAWB, Avicennasis, Awild4, Awild5, AxG, Axegod12, AxelBoldt, Axt, Ayle001, Ayrton Prost, Azov, AzseicsoK, BD2412, BGManofID, Bachrach44, Backspace, BadHabit86, Bahavd Gita, Bakerq, Baldhur, Ballestrain, Bambuway, BanyanTree, Barbe Igor, Baristarim, Baronnet, Bart133, Basawala, Bazonka, Bbryce1, Beach Coma, Beagel, Beatle Fab Four, Beaumain, Beeblebrox, Beenhj, Beeswaxcandle, Beetstra, Bejnar, Belligero, Ben-Velvel, Bender235, Bendono, Benton287, Benzach22, Beregond5, Berkunt, Bernard the Varanid, BernardTom, Bert Schlossberg, Bertie, Bertyrex, Bestalex, Bestsemesterinfo, Beyond silence, Bezenek, Bhadani, Bhargav123, Bhb1991, Bibliomaniac15, Big Moira, Big iron, Biggerj1, Bigthu77, Bility, BillSMASH, Billlion, Billybill, Billybob68, Billybobfro, Binokoron, BirdValiant, Birlanady, Biruitorul, Bjelleklang, Bjmarfito, Bkell, Bkonrad, Bkwillwm, Blackjack48, Blamethrower249, Blaylockjam10, Blethering Scot, Blindman shady, Blizzardstep0, Blood Red Sandman, Bloodfinger, Blue william, Bluecheeseboy, Blueconno, Bluehat66, Bmicomp, Bnwwf91, Bob rulz, Bob999991, BobDively, Bobanni, Bobblewik, Bobianite, Bobo192, Bobrayner, Boenrace, Bogdan, Bogdangiusca, Boing! said Zebedee, Bolonium, Bonadea, Bonde, Bongwarrior, Bonheir, Borisblue, Bornhj, Borovv, Borsi112, Bosonic dressing, Bowei Huang 2, Br'er Rabbit, Bradish, Brandmeister, Brandmeister (old), BrendelSignature, Brewmanz, Brian0918, Brianski, Brichcja, Bridgetansel, BrigadierGeneralBaptisms, Brion VIBBER, BritishWatcher, Brutaldeluxe, Bry9000, Bsdlogical, Btmachine333667, Buaidh, BubbleDude22, Buckshot06, Buffadren, BuickCenturyDriver, Bulatych, Bull Market, Bulldog73, Burbrack, Burningliquid, Bushy moustache, Butros, Buttons, Butwhatdoiknow, Bvjrm, C filev, C-b90, C12H22O11, CJ King, CRKingston, CTJames, CVTautomaton, Cactus.man, CalJW, Calbob66, Calliopejen1, Callymoto, Calmer Waters, Calton, CambridgeBayWeather, Camembert, Camw, Can't sleep, clown will eat me, Canadian-Bacon, CanadianLinuxUser, Canderson7, Caniago, CanisRufus, Canley, Canterbury Tail, Cantus, Capitalistroadster, Captain, CaptainFugu, Caratinga8, CaritasUbi, Carlodellora, Carn, Carnby, Casliber, Cassini83, Catgut, Ccarroll, Cccccccccc, Cchow2, Cdc, Cedrus-Libani, CeeWhy2, Celestra, Certh, Cezzvil, Cfeet77, Cgoodell, ChKa, Chairboy, Chairman S., Chandler Fleck, Chaplin62, Char6128, Charles Essie, CharlesMoscow, Charliemeyers, Chase me ladies, I'm the Cavalry, Chauncyo9, Chavash, Che829, Check two you, Cheese007, Chemical echo, Cherkash, Cherri65, Chochopk, Choop, Chowbok, Chris Roy, Chris the speller, Chris53516, ChrisO, Chrisconey, Chrisfarley1sfat, Chrishmt0423, Chrislk02, Chrism, Chrisminter, Christ2death, Christian List, Christopher Parham, Chtovimenitebemojem, Chun-hian, Chunky Rice, Chzz, Cienade, Cirt, City of Destruction, Civil Engineer III, Ckatz, Claidheamohmor, Clarince63, Clawed, Clcsfist, Cmapm, Cms479, Cod1337, CodeMonk, Codetiger, Coelacan, Colchicum, ColdWarCharlie, Coldsnort, Colipon, Coloane, Colonies Chris, Commander Sergei Bjarkhov, Commi9000, CommonsDelinker, Comp25, Computerjoe, Conman666, Connormah, Conscious, Constantijn, Constantzeanu, Conte di Cavour, Conversion script, Cooldude7273, Cor1314, Coredesat, Corticopia, Cottelli, Cousinjamby, Cptnono, Crazycomputers, Credema, Cremepuff222, Crism, Crit222, Croc22, Crocodealer, Croperz, Crotalus horridus, Cruzian, Cs-wolves, CsDix, Curps, Curriec69, CyberSkull, Cyblue, CyclePat, Cyclonenim, Cylik, CyrilB, Cyrius, CzarB, D-star22, D0t, D4niel11, D6, DARTH SIDIOUS 2, DCGeist, DCowhig, DDima, DJ Clayworth, DO'Neil, DTOx, DVD R W, DVoit, Daa89563, Dabigjc, Daborhe, Dacrone, Dadaist6174, Daduzi, DaemonischEngel, Dalf, Dalta, Damien solta, Daminstant, Dan D. Ric, Dan56, DanKeshet, Dancer6, Daniel5127, DanielCD, Danielkueh, Danieltiger45, Daniil naumoff, Danio, Bibliophylax, Danlaycock, Danlina, Danny, Dantheman531, Dar-Ape, Daraheni, Daremyth, Dark Shikari, DarkElrad, Darkmasterjoey, Darknshadow, Darkspots, Darkwarden, Darkweasel94, Darrendeng, Darthgriz98, Darwinek, Davandron, Davemcarlson, Davenbelle, Daveswagon, David Falcon, David Gerard, David Johnson, David Jordan, David Kernow, David.Mestel, Dbachmann, Dbenbenn, Dcooper, DeServe77, Deacon of Pndapetzim, Deavenger, Debresser, Decman800, Def43, Deflective, Deg2800, Delirium, Delldot, Delxx, Demerzel, Demiurge, Demmy, Dendodge, DenisLG, DenisRS, Denniss, Deon, Der Golem, Descendall, Desiphral, Detnchris14, Deviathan, Dewritech, Dfhjhjjftksfbgr, Dhruvgupta97, Dialog, Didactohedron, Dief88, Diegom809, DigitalGhost, Digwuren, DimaY2K, Dimmy Simmy, Dina, Dino, Dirak, DirkvdM, Discospinster, Djcartwright, Djr13, Dk1panther, Dlrohrer2003, Dlyons493, Dmerthe, Dmeyler, Dmismir, Dmitri, Dmitri Lytov, DmitryKo, Dmn, DoRD, Doc glasgow, DocSigma, DocWatson42, Docu, Documentinghalloween, Dodiad, Dojarca, Dominik92, Donut2, Doopdoop, Dori, Doubleodd, Doug Johnson, DougRWms, Download, Downstrike, Dowsiewuwu, Dppowell, Dputig07, DrJamesMarcus, Draeco, Dralwik, Dramartistic, Drappel, Drbug, Dren55, DriveMySol, Dropbows90, Dru of Id, Dsazonov, Dsp13, Dudtz, Duncharris, Duroy, Dustimagic, Dwarf Kirlston, Dwayne, Dwo, Dwrcan, Dyadyavasya, Dycedarg, Dysepsion, Dysprosia, Dzenanz, Dzied Bulbash, Dÿrlegur, E Pluribus Anthony, E0steven, EJF, ESkog, Eaefremov, Eagle4000, Easybreezy1023, Eazymofuzer, Ecad93, EconomistBR, EddyVadim, Eden87, Editor88, EdoDodo, Edoroth, Edward, EdwardLane, Eewild, Efghij, Ehsan1981, Eightofnine, Eisnel, Ekren, El C, El Snubbe, El Suizo, ElKevbo, Electionworld, Elementanton, Elk Salmon, Elliegurl11, Elmo70, Elockid, Eloquence, Elostirion, Emax, Emberstone, Emery2983, Emigrant123, Emilfarb, Emoelvis, EncephalonSeven, EncycloPetey, Ender78, EngGerm12, Enkyo2, Enormousdude, Enviroboy, Epbr123, EpiC, Epicbobman10, Epicgenius, Eric-Wester, Erik9, Esimal, Esn, Estoy Aquí, Estropes, Eszett, Ethanisasbeancool, Ethelred Cyning, Eu.stefan, Eugrus, Euoa, Eurocopter, Evandroroncatto, Evangelion9490, Everton, Everyking, Evil Monkey, Evilhairyhamster, Ewlyahoocom, Excirial, Explicit, Ezhiki, Ezra Wax, FCSundae, FF2010, Facts707, Fair Deal, Faithlessthewonderboy, Fakirbakir, Falcon8765, Fame, Fan-1967, Fang Aili, Fasgab, Fatal exception, Fatslob, Fedallah, Feedmecereal, FeelSunny, Feeltherbbrducki, Feezo, Felipe Menegaz, FellGleaming, Fellix, Fernirm, Feydey, FidelFair, Fieldday-sunday, Fil usa, Filipdr, Finlay, Fireaxe888, Firsak, Firsfron, Fiscodrawing, FisherQueen, Fisss, Fist of Glory, Fitnr, Fiveyearplan, Flakybizcuit, Flamarande, Flashflash;, Flatterworld, Flexx91872, Flix11, Flowonthego, Flrn, Flubeca, Flying Llamas, Folantin, Football1305, Fram, Frances craig, Francis Tyers, FrancisTyers, Frank Russian, Freakofnurture, Freddie42, Freddiem, Fredrik, Freivolk, FreplySpang, FreshBulletTime, FreshFruitsRule, Fresher, FreyasCrystalizedAngels, Freyr, Frietjes, Frito-lay12345, Frostbitn, Funandtrvl, Funnybunny, Funnyhat, Furrykef, Future Perfect at Sunrise, Fvw, Fys, G. Campbell, G8summit, GHe, Gabbe, Gaius Octavius Princeps, Galati, Galaxydog2000, Galoubet, Gamalon Idogyr, Gamingexpert, Gandygatt, Garik, Garret Beaumain, Garth of NEaB, Gary D, Garzo, Gavrilov, Gaylord22, Gaz, Gdo01, Gearfordeath, Gelovotoz, Gene s, General Eisenhower, General Grievous, Geni, Geoffrie, Geoffspear, Geohem, GeorgeLouis, GerardM, Gertjan R., Gggh, Ghirlandajo, Gibson Flying V, Gidonb, Gigety gigety goo, Gio11211, Girdi, Gjd001, GlaDooo, Glen, Glenn, Globalstatus, Glossologist, Gnomz007, GoOdCoNtEnT, Gogo Dodo, Gohho, GoingBatty, Golbez, Gold heart, Golfcourseairhorn, Good Olfactory, GordyB, Gosso, Gostanford8, Gr8opinionater, GraemeL, Grafen, Graham87, GrahamColm, Grapeme, Graphic, Green Giant, GreenUpGreenOut, Greenshed, GregAsche, GregorB, Gregorik, GrelkJem15, Grendelkhan, Greswik, Grey Wanderer, Greyhood, Griffinofwales, Gringo300, Grosplant, Ground Zero, Grunt, Grye, Gryffindor, Grúnt, Gscshoyru, Guilherme Paula, Gump Stump, Guppie, Guy Peters, Gwdr500, Gzornenplatz, H-ko, HIDECCHI001, HJ32, HJV, Haaqfun, Hadal, Hadiyazzi, Hadseys, Hagedis, Hainzey, Hajor, Half price, Halmstad, HammerHeadHuman, Hanacy, HansHermans, HanzoHattori, Happyme22, Hari6389, Harmil, Harp, Harryboyles, Harveyqs, HasanFerdinand, Hazhk, HaŋaRoa, Hbdragon88, Hdt83, Heff01, Hellerick, Hello32020, HelloAll12, Hellogeoff, Helsinkicommission, Hemanshu, Henk65, Henry W. Schmitt, HenryLi, Heron, Hertok, Herut, Hfsulliv, Highfields, Highvale, Hirudo, HisSpaceResearch, Hisagi, Hiyaninja, Hmains, Hmusseau, HobbesLeviathan, Hodja Nasreddin, HoldenV8, Hollowpens, Holod, Homagetocatalonia, Homero simpstone, Hongshi, Honshuzen, Horlo, Horn123, Horvat Den, HotXRock, Hottentot, Howcheng, Hto9950, Hullaballoo Wolfowitz, Hungaria777, Husond, Hut 8.5, Huw Powell, Hvn0413, Hydriz, Hydrogen Iodide, Hz.tiang, I63G, IGod, IJzeren Jan, IRelayer, IVRacing, Ialsoagree, Iampatel, Ian1981, Ianneub, Iatomm, Icairns, IcePuckScore, Icseaturtles, Idaltu, Idet.proverka, Ief, Ienjoyeating, Ifeldman84, Ifnord, Ifyoudontlikethisthenpleasedonttellme, Igiffin, Ignatpakhotin, IgorMagic, Igordebraga, Iheartjesusveryverymuch, Iketsi, Ikh, Ikiroid, Iksisy, Illegitimate Barrister, Illythr, Iluvjazz, Ilya1166, IlyaMart, Ilyusha V. Novikov, Ilyushka88, Impala2009, Imperatorjan, Improv, Incidious, Indisciplined, Indon, IngSoc BigBrother, Inglok, Ingolfson, Inhakito, Innab, Insuranze, Int21h, Intelligentsium, Inter, Intothewoods29, Introvert, InvisibleK, Inwind, Ioakinf, Ioannes Tzimiskes, Ionut.shaggy, Iridescent, IronPhoenix, Irpen, Irrypride, Isaac Rabinovitch, Ischorr, Islamablad, Istartfires, Italiano111, Itinerant1, Iv1607, Ivan Volodin, Ivan2007, Ivanip, Ixfd64, Izax143, Izzedine, J Di, J Milburn, J budissin, J.K Rowling, J.delanoy, JD79, JForget, JHMM13, JIP, JJARichardson, JNW, JRHorse, JSquish, JTBX, JWSchmidt, Jacek Kendysz, Jack Silver1000, Jack Upland, Jackfield, JackofDiamonds1, JackofOz, Jacksaywhat, Jacoplane, Jafeluv, Jagged 85, Jam1993, JamesBWatson, Jamesday, Jamesinderbyshire, Jamesooders, JamieS93, Jankaspar, Janneman, Janviermichelle, Jarble, Jarkeld, Jaronnelson, Jason Recliner, Esq., Jatrius, Java13690, Jaxl, Jaydlewis, Jayron32, Jbl1970, Jcaragonv, Jcrook1987, JdeJ, Jdforrester, Jebba, Jeendan, JegaPRIME, Jennavecia, Jennica, Jensboot, Jerome Charles Potts, JerryFriedman, Jerzy, Jevansen, Jez, Jezhotwells, Jh51681, Jhendin, Jiang, Jim Douglas, Jimp, Jimtaip, Jirka.h23, Jj137, Jjhcap99, Jjkkl, Jkaplan, Jniech, JoSePh, Joachim Weckermann, JoanneB, JodyB, JoeSmack, Joefu, Johan Elisson, John, John Price, John Quincy Adding Machine, John Smith M.D., Ph.D, John of Reading, John254, JohnG62, JohnWittle, Johndyra, Johnhenry312, Johnleemk, Johnluisocasio, JohnnyWiki, Jojolias, Jomorepinch, Jonatan Swift, JonathanDP81, Jonathunder, Jonearles, Jonjames1986, Jonkerz, Jose77, Joseph Solis in Australia, Joshualoljosh, Jossi, Jovianeye, Joy, Joyous!, Jozdnog, Jpeob, Jpgordon, Jph, Jpople, Jrockley, Jtkiefer, Jugbo, Julekmen, Jumbuck, Jung dalglish, Jusdafax, Jusjih, Just zis Guy, you know?, Justinjuicebox, Jvanore68, Jwissick, K a r n a, K.O.T., KNewman, KTC, Kablammo, Kaiser matias, Kajervi, Kalsrag, Kanchelskis-it, KapilTagore, Karabinier, Karesu12340, Karl Marx, Kascnef82, Kassabov, Katydidit, Kayac1971, Kaz, Kazak, Kazayta, Kbh3rd, Kd4nuh, Kda2011, Keegan, Keelm, Kefeer, Keilana, KeithB, KelleyCook, Kelstonian, Keluar94, Kelvinc, Kemor, Kenmore, Kev2310, Kevin B12, Kgrad, Khashyar, Khatru2, Khazar2, Khelnor, Khoikhoi, Khvalamde, Kids4Fun, Kikichugirl, Kilhan, Kilimac, KillaShark, King Lopez, King Zebu,

Kingpenguins2007, Kingpin13, Kintetsubuffalo, Kkkdc, Kkm010, Klenje, Kman665, Kmorozov, Kneiphof, Knutux, Koavf, Kolja21, Kolkhian, Koov, Kosebamse, Kotniski, Koyaanis Qatsi, Kozuch, Kpalion, KrakatoaKatie, Kralizec!, Krawndawg, KrayCeremoush, Krenakarore, Krich, Kristallstadt, Kross, Krukowski, Ktoto89, Kuban kazak, Kubigulo, Kuiavian, Kuimov, Kukini, Kulikovsky, Kungfuadam, Kungming2, Kurieeto, Kurtber, Kuru, Kurykh, Kusma, Kusunose, Kuzmaxster, Kwamikagami, Kwertii, Kwesadilo, Kyle Barbour, L Kensington, LEGOBRICKSTER3, LMANSH, LUCPOL, La goutte de pluie, Lacrimosus, Lactasamir, LadnaV, LahDeeDah7, Laikas, LairepoNite, Lakers, Lantern40, Lappip, Lapsed Pacifist, Le Anh-Huy, Leadgold, Leandrod, Leapsorymt, Lear's Fool, Lectonar, Lenev, Lennonist, Leolaursen, Levelistchampion, Levineps, Lexa87, Lexicon, Liamdaly620, Library time, Life is like a box of chocolates, Liftarn, Lightdarkness, Lightmouse, Ligulem, Likelife, Likthem, Lilac Soul, Lilmaniac, Limideen, LincolnSt, Lindenhurst Liberty, Ling.Nut3, Linguist Dan, Linkfan, Linnell, LiquidOcelot24, Little Professor, Livajo, Llamabread, LlywelynII, Lockesdonkey, LokiiT, Lokqs, LonelyMarble, Longhair, LookingGlass, Loonymonkey, Looxix, Lord Vader, Loren36, Lothar von Richthofen, Lotje, Lowe4091, LtPowers, Luisaterzontz, Luisztdt, Luk, Lumen24, Luna Santin, Lunchboxhero, Lupo, Lvivske, Lysy, M.nelson, MC MasterChef, MER-C, MJCdetroit, ML, MONGO, MSGJ, MZMcBride, Mac, Macintosh User, MackSalmon, Mad Max, Maelnuneb, Magicmonster, Magister Mathematicae, MagnusA, Mailer diablo, Majorly, Mak2007, MakmoudHassan, Malarious, Malcolm Farmer, Male1979, Malerin, Malhonen, Malick78, Malmo Blue, Malo, Man vyi, Mangalaiii, Mani1, Manop, Mansmokingacigar, MapsMan, Maravjua, Marc87, Marcika, Marco Neves, Marcus2, Mardus, Marek69, Mareklug, Mariapoliantseva, Mark rico, MarkGallagher, MarkSweep, Marko7, Marlow4, MarphyBlack, MarsRover, Martarius, Martin23230, Martin451, Martinwilke1980, MasterShake1000, Materialscientist, MathMartin, Matreal, Mattbrundage, Mattisse, Mattrulez, Matts computer, Matty, Maunus, Mausy5043, Mav, MaxSem, Maxim, Mazerunner, Maziotis, Mb nl, McDogm, McGeddon, Mcorazao, Meadow Soprano Wheels!, Mediolanum, Medo9, Medvedev40, Meegs, Mefistofele, Meganhernandez, Melikamp, Mememethem, Mentifisto, Merbabu, Merenzon, Merotoker1, Mesgul82, Meursault2004, Mfa fariz, Mfzb04, Mgiganteus1, Mhacdebhandia, Mhardcastle, Mherlihy, Mhocker, Mhsb, Mic, Michael Devore, Michael Frind, Michael Hardy, Michael5046, MichaelMaggs, Michaelas10, Michaelmas1957, Michaelphelps'sbongwater, Michaelwuzthere, Mike Rosoft, Mike by, Mike1024, Mikhail Garouznov, Milkyeyes, Miller17CU94, Mimihitam, Minesweeper, Minimac, Mirv, Misaq Rabab, Misortie, Missionary, Mistakefinder, Misza13, Mitchking23, Mitzu17, Miyokan, Mjpieters, Mobile6681, Modulatum, Modþryð, Moejoeonline, Mohawkjohn, Molly-in-md, Monedula, Monk, Monkeyman, Monterey Bay, Mooo, Moravice, Moreschi, Morgankevinj huggle, Morwen, Moscvitch, Moshe Constantine Hassan Al-Silverburg, Moxfyre, Moxy, Mqe, Mr Bartels, Mr Stephen, Mr. Lefty, Mr. Military, Mr. Yooper, Mr.Unknown19, MrJacky, MrSchmenge, Mrbrush, Mrs Trellis, Mrtwig, Ms2ger, Mschel, Mschiffler, Muhaidib, Muncadunc, Munci, Muntuwandi, Musaabdulrashid, Mushroom, Musical Linguist, Musser, Mustafagheran, Mwanner, MyNameIsNotBob, Myanw, Myunknownside000, Mzajac, N-edits, N-true, N3X15, NHRinger, NIR-Warrior, NPPyzixBlan, Naddy, Nadirali, Nadyes, Nakon, Nanobear, Nanshu, Narking, Narxysus, Naryathegreat, Nat Krause, Natalia O., Natalie Erin, Natgoo, Nathanww, NativeForeigner, Natl1, Naturada137, NawlinWiki, Nayvik, Neartobrum, Neegjeris, Neglen, Nehrams2020, Neil916, Neonmario, NerdyScienceDude, Netoholic, Netol, Neumannk, Neutrality, NeutronTaste, Never give in, NewEnglandYankee, Newsrussia, Ngyikp, Niceguyedc, Nichalp, Nick Number, Nick mallory, Nick-D, Nick.mon, Nick125, Nickshanks, Nicksss93, Nicksukh, Nickx33, Nickydude, Nicolas Love, Nigholith, Nightstallion, Nihil novi, Nihiltres, Nikai, Nikitadanilov, Nikkimaria, Nikola Smolenski, Nirvana888, Nissi Kim, Nivix, Nixer, Nizzil, Njaelkies Lea, Njeman, Njoedits, Nlight2, Nmpenguin, No Guru, No limits99, No.13, Noah Salzman, Nobleeagle, Noiseidea123, Noisettes, Nojustno, Nono64, Noodlenut, Noommos, Noozgroop, NorthernThunder, Northumbrian, Notedgrant, Notheruser, Notthe600, Nsk92, NuclearVacuum, NuclearWarfare, Nug, Numbo3, Nv8200p, Nyttend, Nérostrateur, OBdude7, Obradovic Goran, Observerq, Occamy, Octane, Ogatai, Oghmoir, Ogmo, Ohconfucius, Ohmygod766, Ohnoitsjamie, Okapi, Okolobaxa, Okome, Olegthezenon, Olegwiki, Olgasavina, Olgerd, Oliphaunt, Olivier, Ollivier, Olna, Olorin28, Olvegg, Omega360, Omicronpersei8, Ominpanda, OneGuy, Onehundredyears, Only Red, Onopearls, Onorem, Openrussia, Opodo123, Optigan, Orangerider, Orenburg1, Orionist, Orthuberra, Ostap R, Otolemur crassicaudatus, Ouip, OverlordQ, Overnight darkness, OwenBlacker, OwenX, Owned4itall, Oxana879, PANONIAN, PFHLai, PIO, PRH, PSzalapski, PTYLER22, Pablo X, PaddyBriggs, Paine Ellsworth, Pajfarmor, Pakhomovru, Palefire, Pama73, Panairjdde, Papa Carlo, Parhamr, Parishan, ParisianBlade, Pass a Method, PasswordUsername, Pathoschild, Patrick, Patrick Berry, Paukrus, Paul August, Paul Benjamin Austin, PaulGarner, PaulHanson, Paulinho28, Paulscho, Pavel Vozenilek, Pax, Pax:Vobiscum, Pb30, Pearle, Pedroshin, Pegasus1138, Pensionero, Penubag, Perceval, Pereleshin, Permico, Perovic, Persian Poet Gal, Peru-Russian888, Peruvianllama, Peter Horn, Peter Isotalo, PeterPredator, PeterisP, Pethr, Petri Krohn, Peyman Ghasemi, Pgan002, Pgk, Phaedriel, Philip Trueman, Philipmj24, Phillip J, Philpm930, Pholi, PiMaster3, Pianist ru, Picaroon, PickupmanUP1966, Piepoop, Pigman, Pilcrow, Pilotguy, Pinball22, PinchasC, Pinethicket, Piotr9, Piotrus, Pip92, Piratedan, PizzaBox, Plapre09, PlatypeanArchcow, Pletet, Pmanderson, Poetaris, Poker face20, Polaron, Polly, Polmandc, Polnian, Polon11, Polsequ95, Poor Yorick, Poptartbeebee, Porkies84, Postdlf, Pouya, Pras, Prawn3944, Princesin72, Principia1, Profmajom, Pronoun, Prospectmira, ProtectionGate, Pseudomonas, Psy guy, Ptorchilo, Ptrt, Pudeo, Purpleturple, Pygenot, Pyroclastic, Python eggs, Pz821, Q43, QFlux, QZXA2, Qasaqsuyu, Qatter, Quandapanda, Quarkington, QueenCake, Queenmomcat, Qutezuce, Quuxplusone, Qwertzy, Qxz, R'n'B, R-41, R7K, RJN, RK, ROSARIO STEPHEN, RPIRED, Rab V, Rabindra Baral, RadicalBender, RadioKirk, RafaAzevedo, Rahlgd, Rail88, RainWingWolf, Raja Hussain, Ral315, Ramand, RamblinWreck3, Ramdrake, Rameriz, Ran, Random contributor, RandomP, Randomblue, Ranveig, Rarelibra, RaseaC, RashersTierney, Rasmus Faber, RasputinAXP, Raven4x4x, Raymond arritt, Razionale, Rclark53, Rd232, Really Spooky, Recker, Reconsider the static, Red Director, Reddi, Reddragon555632, Redking7, Redstarsldr, Redvers, Regionalsimp, Rehnn83, Reichenbach, Remisc, RenamedUser01302013, Reticulated, Retired username, RetiredWikipedian789, Rettetast, Rev4ricky, RexNL, Rezashah4, Reznoff4, Rhysn, Riana, Rich Farmbrough, Richard David Ramsey, Richard0612, RichiH, Rick Block, Rick DeLong, RickK, Ricky81682, Ricky@36, Rickyrab, Rivertorch, Rjcflyer@aol.com, Rje, Rjensen, Rjwilmsi, Rmcubed, RoadTrain, Robert K S, Robert Merkel, RobertG, RobertLunaIII, Robertgreer, Robertp72, Robinkruyt, Robwingfield, RockMFR, Rockador, Rocket71048576, Rodasmith, Roentgenium111, RohirrimRider, Romanius, Romanm, Rome1453, Ronbo76, Rory096, RossPatterson, Rossen3, Rowe 13, Roy da Vinci, RoyBoy, Royalguard11, Rrburke, Rreagan007, Rrtycoon, Ruhrjung, RuineR, Runningfridgesrule, Ruok, Rusish, Russavia, Russian Luxembourger, Russian-Reaper, Russiansoul, Rwxrwxrwx, RxS, RyanGerbil10, Ryanforesight, Ryulong, S.B. Odin, S119234, S5switch, SCJohnson77, SD6-Agent, SDC, SECProto, SGBailey, SGGH, SJK, SMC, SOPHIAN, SPORTYGRL012, SQGibbon, SS451, STARWARSGEEK, SWO1975, Saffloped, Sagaciousuk, Saganaga, Saiga12, Saluyot, Sam Korn, Sam Weber, Samantha555, SamdupLa, SamdupLama, Samir, Samrolken, Samuel Blanning, Samuelsen, Samwb123, Samy511, Sanbec, Sandip90, Sandlas Juagas, Sandstein, SandyGeorgia, Sango123, Sardanaphalus, Sarranduin, Sasa2007, SashaGolikov, Sasquatch, SatnSpwn, Satori Son, Sayden, Sca, SchfiftyThree, SchmuckyTheCat, Schnellundleicht, SchnitzelMannGreek, Schooks59, Schoolthing, SchuminWeb, Scientistmohamed, Scipius, Sciurinæ, Scott Burley, Scottwilleke, Scrazen, Scythian1, SeNeKa, Sean K, Secfan, Seicer, SelfQ, Selfworm, Seniorfox, Senzangakhona, Sergey.belyakov@gmail.com, Seryo93, Sesu Prime, Seth harsch, Sethmahoney, Sevela.p, Sevenlee, Seyrüsefer, Sfahey, Shadiac, Shadow Android, Shadow1, Shadowjams, Shadowxfox, Shak-74M, Shamir1, Shanes, Shannon bohle, Shattered, Shattered Wikiglass, Shisock, Shiva Evolved, Shizhao, Shoaler, Shogunpk, Shredder2012, Shultz IV, SiPlus, Signalhead, Signsolid, Silly rabbit, SimonP, SimonX, Simonsaysabc123, Sindri, Sir Edgar, Sir Nicholas de Mimsy-Porpington, Sirgregmac, Sirius85, Size J Battery, Sjakkalle, Skjorgen, Sky Attacker, Skybon, Slakr, Slashme, Slif3r, SlimVirgin, Slomox, Slowking Man, Slurpy121, Slyguy, Smallbones, Smalltime0, Smarkflea, Smashing, Smb1001, Smeira, SmthManly, Snaildaddy, Snailwalker, Snap Davies, SndrAndrss, Snowolfd4, SoLando, SoSaysChappy, SoWhy, Sociallyawkwardpenguin, Soetermans, Sokol 92, Solentways, SolidEdgeGuRu, Solitaryxchild, SomamcFloater, SomeFreakOnTheInternet, SomeStranger, Something20, Somethingsometing, Son of thunder, Sonicsuns, Soosed, Sophus Bie, Sour cream monster8, Sozsoz, Sp33dyphil, Spaceboy492, Spangineer, Spartan65116, SpartanG117, Spartaz, SpecMode, Spellcast, Spespatriae, Spettro9, SpigotMap, Spike35031, Spliffy, SpookyMulder, Spot87, Spotty11222, Spring01, SqueakBox, Squeezeweasel, Squiddy, Squish2013, Srushe, Sshirokov, Ssolbergj, Ssr, St.daniel, Stalin88, Stan Shebs, Standfest, Starnestommy, Starwars10, Stassats, Staszek Lem, Staxringold, Stealthbreed, Steel, Steinbach, Stella Vesper, Stephen MUFC, Stephenb, Stephreef, Steve2011, Steven J. Anderson, Stevenmitchell, Stevenochs, Stitchix, Stoljaroff1987, Storkk, Student7, Subtropical-man, Suede, Suffieldlax24, Suffusion of Yellow, Suigntou11011111, Suklaa, Sungjwoo, Sunray, Super-Magician, SuperDeng, Superdude602, Superking, Superninjakatmanofeditorialninjakats, Supertask, Superzohar, Supreme Deliciousness, Surferus, Surlyduff50, Svmich, Sweden555, Swikid, Swimm2, SwinSvinoza, Swu1992, TAG-A-b10, TFCforever, TKD, TUF-KAT, Taamu, Tajik24, TakeMemoriesAway, Tamarav, Tango, Tarret, TarzanASG, Tasc, TastyCakes, Tat1642, Tataral, Tauiris, Tavork, Tavrian, Taw, Tb, Tbhotch, Tealdon, Teamweeeman, Techman224, TedE, Teevek, Teh1337indian, Tennisuser123, Teo jun min, Terailer, Terence, Tesi1700, Testu, TexasAndroid, Texmon, Texture, Tgv8925, Thatswhatsup, The Anome, The Banner, The Epic Hawk, The Green Fish, The Rambling Man, The Thing That Should Not Be, The Tonya, The Transhumanist, The pencil fudge, The tooth, The71, TheAnimalSkull, TheChanger, TheDJ, TheLeopard, TheMadBaron, TheNewPhobia, TheRanger, Thecrazydonvingigante, Theda, Thedcm, Thedefenceman, Thedjatclubrock, Theguyotc, Thejadefalcon, Themightyquill, Thenowhereman, Theprophetmyass, Therequiembellishere, Therexbanner, Theseven7, Thesevenseas, Thryduulf, Thue, Thunderrat34, Tide rolls, TigerShark, Tigerhawk47, Tim Starling, TimBentley, TimComm, Timberlax, TimothyPilgrim, Timrollpickering, Tinsue, Tired time, Tisquantum, Titoxd, Tiwonk, Tjl1128, Tobby72, Toccata quarta, TomStar81, Tombrider1, Tombseye, Tomcat7, Tomjoneser, Tonik, Tony1, Tonyfaull, Topgun205, Topuria, TortoiseWrath, Tovarish14, Toy111, Tpbradbury, Tra, Trademarx, TransUtopian, Transilvanian Dracula, Trasman, Treveraritz, Trevor MacInnis, TrollingTrollerTroll, TrumanRu, Trusilver, Trust Is All You Need, Tsirk13, Tsob, Ttony21, Tulandro, Tulkolahten, Turgeis, Turgidson, Turkeli, Turlo Lomon, Tvdog, Twillisjr, Ty1993, Typhoonchaser, TyrocP, Tyw7, U.Steele, Ubiquinoid, Ucucha, Ufwuct, Ugen64, Ugur Basak, Ulises Heureaux, UltimatePizzaLord, Uncle Dick, Uncle Scrooge, Underlying lk, Underorbit, UnicornTapestry, Unionin, Universe=atom, Unsc, Unschool, Untifler, Unused0024, Urfinze, Uris, Urod, Useight, User BartS, User27091, User6854, User86654, Usergreatpower, Utcursch, Uuu87, V4711, VORON SPb, VVPushkin, Valentinian, Valentino, Valio subaru, Valoradam, Van der Hoorn, Van helsing, Vancouverguy, Vaniba12, Vanished user 05, Vanished user 9i39j3, Vanished user aqpoi4u3tijsrfi, Vardion, Varnav, Vary, Vas123, Vega84, Vegaswikian, Vegidio, Venajaguardian, Ventur, Verdadero, Verseruns, Vetrov, Vice regent, Viciousvogel, Victor12, Victoria2007, Victory in Germany, Vidokov, Virgule82, Vlad Tepes IV, VladPutin, Vladdacomrade, VladimirKorablin, Vlasds, Vlastoff, Vmenkov, Vmrgrsergr, VolatileChemical, Volodin, Volodyanazarkevych, Volunteer Marek, Vortexica, Voyevoda, Vrghs jacob, Vrinan, Vroman, VsevolodKrolikov, Vsmith, Vssun, Vuo, Väterchen, WGee, WLFV, Wafulz, Wallie, Wanjuscha, Ward3001, Warfvinge, Wassamatta, Wassermann, Watashiwabakayo, Waterhoof, Wavelength, Wavesswung99, Wayne317, Wayward, Wecameasromans, Weedro, Welsh, Wenli, Wesha, WesleyDodds, Wetnose23, WhereAmI, Whichtrial, WhisperToMe, WhitmanNE, Who, Wicojrpr, Wiglaf, Wik, WikHead, Wiki alf, Wiki1609, WikiLaurent, Wikiagoo, Wikibojopayne, Wikiborg, Wikieditor06, Wikignome1213, Wild wadi, Wildyoda, Wilfried Derksen, William Allen Simpson, William Avery, Willqualey, Wimt, Windman233, Wisco, Wjejskenewr, Wjfox2005, Wk muriithi, Wknight94, Wliiam·Shakespeare, Wollont, WoodElf, Woohookitty, Worldtimezone, Worldtraveller, Wrath of Hell, Wsiahoffman, Wælgæst wæfre, X!, X96lee15, XAKxRUSx, Xahm2307, Xalpharis, Xdenizen, Xerstau, Xezbeth, Xfactor2000, Xiner, Xoloz, Xyzzyva, YSSYguy, YUL89YYZ, Yah, Trick, Yah, Yahel Guhan, Yamamoto Ichiro, Yanksox, Yartch, Yasis, Yassou Jayne, Yath, Yattum, Yeaboiitskurt, YellowMonkey, Ymblanter, Yonas29, Yoooooooooork, Yopie, Yorkshirian, Youmps, Youramelon, Yozer1, Yt95, Yuckfoo, Yulia Romero, Yurko57, Yusha0450, ZRetro, ZZeta, Zaf, Zaharous, Zakiya alaamri82, Zandweb, Zaphrost, Zarniwoot, Zealander, Zeanrivera, Zedshort, Zeliboba7, Zenit4ever, Zepheriah, ZincOrbie, Zlerman, Zloyvolsheb, Zmaher, Znayka, Zoe, Zoeneo480144, Zoloto-Russland, Zombiecapper, Zondor, Zscout370, Zserghei, Zsinj, Zurkhardo, Zvar, Zz pot, Zzuuzz, Zzyzx11, Zé da Silva, Мистер ОКО, РКП, Скампецкий, Эрманарих, דוד, مرساب, 3291 anonymous edits

Krai *Source*: http://en.wikipedia.org/w/index.php?title=Krai *Contributors*: Ajh1492, Aleksandr Grigoryev, Alexander Sokolov, Alfanje, Altenmann, Andres, Artegis, BaF, Bill3000, Cautious, Chakat Rimm, Danrok, David Kernow, Drbreznjev, Ev, Ezhiki, Gilliam, Grutness, Hercule, Heron, Iguacu, Juro, Lazar Taxon, Lokys dar Vienas, MAXXX-309, Mani1, Marimarina, Mattisse, Mav, Milonica, MrPMonday, Nixer, NoPuzzleStranger, Olivier, Pavel Vozenilek, Petri Krohn, R9tgokunks, Riurik, Rune X2, SchreiberBike, Stephensuleeman, TastyPoutine, Tobias Conradi, TurkChan, V1adis1av, Valip, Vmenkov, Vuong Ngan Ha, Yubcvan, 26 anonymous edits

Subdivisions_of_Russia *Source*: http://en.wikipedia.org/w/index.php?title=Subdivisions_of_Russia *Contributors*: 102orion, A4, Ahoerstemeier, Albania T, Altenmann, Anthony Appleyard, Apoivre, Aris Katsaris, Beyond My Ken, BilCat, Bjankuloski06en, Brianski, BrownHairedGirl, Chanheigeorge, Crusoe8181, Danny, David Kernow, Dimitris, Dl2000, Docu, DonAvero,

Electionworld, Enviroboy, Everyking, Ezhiki, Fabartus, Firespeaker, Flrn, Frysun, GregorB, Greyhood, Hibernian, Hodja Nasreddin, Hvn0413, IZAK, Isthmus, Ivan Volodin, Jao, Joy, Just Another Dan, Khoikhoi, Kohka, Kokoo, Kostisl, Kpalion, Li-sung, Mattisse, Mic, Monedula, Morwen, Muriel Gottrop, Mzajac, New World Man, Olivier, Orenburg1, Orthuberra, Patrick, Paukrus, Pmmollet, Ran, RandomCritic, Rarelibra, Rredwell, Russavia, SPUI, Sardanaphalus, Scipius, Seivad, Smooth O, Superzohar, TarzanASG, Tavork, The Transhumanist, Thumperward, Tiraios-of-Characene, Tobias Conradi, TrueColour, Untifler, Vanjagenije, Woohookitty, Ymblanter, Zumbo, Милан Јелисавчић, РКП, 47 anonymous edits

Administrative_centre *Source*: http://en.wikipedia.org/w/index.php?title=Administrative_centre *Contributors*: Aleksandr Grigoryev, Applefilling, Bhny, Breadandcheese, Charvex, Cncs wikipedia, DDima, Discospinster, Ezhiki, Hectorthebat, Hgb1217, Intangible, JeremyA, Jza84, MRSC, Page Up, Pilim, Pol098, Retired username, Thorsen, Tomasino96, Untifler, Whoatherekiddo, 3 anonymous edits

Russian_Census_(2002) *Source*: http://en.wikipedia.org/w/index.php?title=Russian_Census_%282002%29 *Contributors*: Alex1709, Altenmann, Berganus, Bigbluefish, Biruitorul, Chenxlee, Colonies Chris, Conscious, Cybercobra, Dbachmann, Dewritech, Ezhiki, Greyhood, Hardscarf, Heallo88, Jevansen, Joseph Solis in Australia, Khoikhoi, Koavf, Kransky, MER-C, Marmale, O'Dea, Shatter Resistance, Tim!, Valip, Vmenkov, 16 anonymous edits

Russian_Census_(2010) *Source*: http://en.wikipedia.org/w/index.php?title=Russian_Census_%282010%29 *Contributors*: Ardsarea, BD2412, CommonsDelinker, DonaldDuck, EdJohnston, El Snubbe, Ezhiki, Greyhood, Jedibob5, Jiujitsuguy, LokiiT, Lvivske, Maxim11maxim, Schekinov Alexey Victorovich, Serjio-pt, Shatter Resistance, Spb1605, Toccata quarta, Викимонетчик, 3 anonymous edits

Chuvash_people *Source*: http://en.wikipedia.org/w/index.php?title=Chuvash_people *Contributors*: .:Ajvol:., AB, Aldux, Aleksandr Grigoryev, Ali Savatar, Aramgar, Attlmt, Bad worse wikip, Ballestrain, Barefact, Bezuidenhout, Cezzvil, Chapultepec, Chavash, ChrisO, Ciuvasc, Darwinek, Deb, Dirrival, DopefishJustin, Drbreznjev, EdJohnston, Elmju, Epyue, Everyking, Evlekis, Ezhiki, FLAGELLVM DEI, Flying Penguin, Fnr Kllrb, Fulviian, Gilgamesh, Gits (Neo), Gkklein, Glossologist, Greczia, Green Giant, Gyte75, Heqs, Hibernian, HistoryGirl08, Hmains, Hottentot, Hurmata, Icairns, Ida Shaw, Jarble, John of Reading, Joostik, Jorge Stolfi, Joy, Khoikhoi, Kjetil r, Kwamikagami, Languagehat, LiDaobing, MacRusgail, Mak13, Mamaberry11, MaxSem on AWB wheels, Mefisteron, Mike Rosoft, Mttll, Neshmick, Nick Number, Nozdref, Olahus, Omnipaedista, PANONIAN, Pensionero, Ralfins222, RicardoDomingguez, Rjwilmsi, Rudjek, Schwert von Feuer, Sfarea, Shokhruhkuz1, Shrigley, Spring01, Suhardian, Te5, Themastertree, TimBentley, Tkinias, Tobias Conradi, Trigaranus, Ufwuct, Untifler, VVPushkin, Verdlanco, Viktor-viking, Volunteer Marek, Woohookitty, XStandartX, Xyboi, Yalens, YixilTesiphon, Youngamerican, Zahidulla, Zakronian, Регистрироваться, Регистрироваться 00, 163 anonymous edits

Russians *Source*: http://en.wikipedia.org/w/index.php?title=Russians *Contributors*: 3 Löwi, 37rrga, 3Peats, A.b.s, ABF, Aaker, Abb615, Aceshigh123, Acroterion, Adjwilley, Aelfthrytha, Afrasclient, Afuzzyllama, Aherunar, Ahmad2099, Ahnode, Ahosek, Airunp, Aitias, Aivazovsky, Ajh1492, Akamad, Alan16, Aleksei, Alex '05, Alex Bakharev, AlexanderKaras, Alm93, Alt6878ggtg, Altenmann, Amatulic, Amherst99, Amire80, Anandks007, Anaraug, Andonic, Andrew Shane, Andrwsc, Andycjp, Angelo De La Paz, Angstorm, AnimalLover323, Ann Stouter, Anna Lincoln, Antandrus, Anthon.Eff, Antidote, Apoivre, Arcollins44, Arcot, Argentina 678, Ari777il, Arkharov, Art LaPella, Arthur889, Artura56, Asidemes, Asterno, Ateapotist, Atitarev, Auroranorth, Avala, Babafira123, Bacian, Bandurist, Baseball Watcher, Baseball1234, Bavarian aryan, Bejnar, Big Adamsky, BigBlueMonster, Bilby, BlakeCS, Blankfaze, Bluemoose, Boccobrock, Bogdan, Bomac, Bongwarrior, Boodlesthecat, Boris ru, Bovlb, Brandmeister, Brianga, BrokenSegue, BulgarianMacedon, Burschik, Bushroot, C+C, CPT.Powers, CWenger, CalJW, CaliforniaAliBaba, Calliope224, Can't sleep, clown will eat me, CanadianLinuxUser, Captain Obvious and his crime-fighting dog, Captain panda, Casper2k3, Cdecoro, Charles Essie, Choieuna, Chris the speller, Chrisminter, Christian75, Cinik, Cirexer, Clicketyclack, Clod2105, Cocytus, Colchicum, Colipon, Commander Keane, CommonsDelinker, Confuzion, Constantzeanu, Contributor777, Cool3, Corpx, Countdown, CountdownCrispy, Courcelles, Crazycomputers, Cst17, Cubbi, Curps, D, DVdm, DVoit, DaL33T, DabMachine, DalmatinoA, Daniel il, Daniscool6, Darkwind, Darth Panda, Darwinek, David.Mestel, David1010, Dbachmann, DeadEyeArrow, Delirium, Deville, Diannaa, Didactohedron, Digital1, Dirak, Discospinster, DocSigma, DocWatson42, Dojarca, Domnu Goie, Don Alessandro, Don Cuan, Donner60, Download, Dpv, Drbug, Ducknish, Duja, E10ddie, E2eamon, Eadthem, EagerToddler39, Eclipsian, Electro Magnetic Killer Radiation, ElisGitt, Elockid, Emax, Ember of Light, Emilfarb, Epf, Epischedda, Esimal, Explayer1993, Explicit, Ezhiki, Falcon8765, Fartherred, Fat&Happy, Faustian, Fayenatic london, Feedmecereal, Feer, Fezzesarecool, Fisenko, FisherQueen, Flewis, Flying Saucer, FlyingToaster, Fortyfeet, Frank mad, Freedomlinux, Freize1, Fryed-peach, Frze, Funandtrvl, Future Perfect at Sunrise, Fuzheado, Fwouter, Gaius Cornelius, Galassi, Gallus Gallicus, Garik, Gayb, General Grievous, Georodin, Gephart, German1337, Ghirlandajo, Gholam, Gigiman21, Glossologist, Gmaxwell, God of Sins, Gogo Dodo, Gosso, Gr8opinionater, Green Giant, GregLoutsenko, Gregbard, Greyhood, Gudbrud, Guinness man, Gwernol, Hairhorn, Haldraper, Hatchback1, Hburn21, Hdt83, Headbomb, Hede2000, Helilooja, Heonsi, Herbo161291, Hergilei, Hibernian, Hmains, Horvat Den, HotelRoom, Hottentot, Hu, Hullaballoo Wolfowitz, Humanophage, Humus sapiens, Hungaria777, HungarianWarrior92, Hunnjazal, Hut 8.5, I'm sexy, I'm hot, I'm everything your not, I1990k, Ian.thomson, Iceager, Ifeldman84, Ilgiz, Ilikemen2345676, Illythr, Ilveon, Ilya1166, Imasleepviking, Inhakito, Intgr, Irpen, Island, Itai, IvanP, J.delanoy, JNW, Jack who built the house, Jacobjohnson123456789, JamesBWatson, Jameswilson, JamieS93, Jauerback, Jawadreventon, Jb12332, Jeronimo, Jiang, Jim1138, Jncraton, Jobas, John254, JohnKreis, Joseph Solis in Australia, Joy, Jpbarrass, Jpgordon, Jstamos, Jyril, Jönssontvå, K1Bond007, KGasso, Kaganer, Kaiser matias, Karenjc, Kazak, Kaztrinh, Kbdank71, Kbh3rd, Kheo17, Khoikhoi, Kids4Fun, KillaShark, Kittymonk4, Klandhee, Klausness, KnowledgeOfSelf, Koavf, Kober, Kolt, Kostan1, Krisma12234, Kross, Krotx, Ksenon, Ktoste, Kuban kazak, Kukini, Kuohatti, Kusamanic, Kwamikagami, Ladnadruk, Lantean Eric, Laurenhazel, Laveol, Laxxx4lyfe, Le Anh-Huy, LeVoyageur, LeaveSleaves, Lectonar, LeilaniLad, Leutha, Libsys, Lightmouse, LilHelpa, Ling.Nut, Linkinaha, Linnell, Lionenause, Lockesdonkey, Log in, log out, Lokodd, Lori89, Lovethemcchockens, Lovingoni, Lvivske, Lysy, M.V.E.i., M5891, MC10, MER-C, MJO, MK8, MaGioZal, Mall89, Magioladitis, Makecat, Mallocks, Marauder40, Marcus2, Marek69, Markov Vasiliy, Marmalade101, Materialscientist, Mathias2010, Maximaximax, Mazca, Med, Mediran, Melesse, Menchi, Migdejong, MiloKral, Missionary, Miyokan, Moab1111111111, Modulatum, Monkeymonkeypanties, Monterey Bay, Moxy, Mp12166, Mr Tan, Mttll, MuZemike, MusikAnimal, Mysdaao, Mzajac, N-edits, NCurse, Nakon, Nanib, Nascigl, Natural Cut, NawlinWiki, NeoJustin, Nerp65, NewEnglandYankee, Ngriser, Nicgator, Nick Number, NickCT, NickCatal, NickMartin, NightOnEarth, NikitaAlexoivch1, Nikola Smolenski, Nikolairoyale, Nikolosa, Nikulinvitaliy, No Free Nickname Left, Noelmantra, Novidmarana, Ntsimp, Nubiatech, Nymf, Obradovic Goran, Ohnoitsjamie, Olegwiki, Olgerd, Olivier, Olvegg, Omicronpersei8, Onorem, Opinoso, Ostap R, Oxey89, PANONIAN, PMK1, PSUMark2006, PacificWarrior101, Papa November, Parishan, Passwd, Paul August, Paul Barlow, PaxEquilibrium, Pdcook, Pehlakas, Pensionero, PeterPredator, Petri Krohn, Pgan002, Pgk, Pharaoh of the Wizards, Philip Trueman, Pinethicket, Pizzadud, PlatonPskov, Playclever, PocketMoon, Poetaris, PranksterTurtle, Praxis101, PrincessofLlyr, ProhibitOnions, Protagon, Prusak, Psy guy, Pudeo, Pudorovsky, Purpleturple, Puuslik, Qing Yi G, Quantumobserver, Quintote, R'n'B, R1990u, RCRC, Rabbed, Rago, Rajputbhatti, Ralfins222, RandomP, Ranveig, Rapheal99, RasN, Ratemonth, Raven in Orbit, Raymond Cruise, Rekcut, Rettetast, Rex Germanus, RexNL, Rheinguld, Ri hwa won, Rich Farmbrough, Ringm, RiverStyx23, Riwnodennyk, Rjwilmsi, Rndomuser, Roadrunner, Roadrunner11, Robertfreemanfund, RolandR, Romanm, Ronhjones, Roo72, RoseAphro, Rosssucks, Rune X2, Runespace, Runewiki777, Rushton John, Russavia, Russia333, Russia77798, Russian Luxembourger, Russiangangster, SANCEZZZ, SJP, SOPHIAN, Saddhiyama, Saga City, Sally101, Salmonish, Samachablo, Sand Patrol, Sanya3, Sasha l, Sassf, Sauceysauce, Scaife, ScaldingHotSoup, SchreiberBike, Sciurinæ, Sco85, Scootey, Scythian1, Sedatx41, Several Pending, Sfmammamia, Shamir1, Shanes, Shanesterman, ShelbyNewsome456, Sheridan, Shpakovich, Shrimp3000, Skanter, Slaja, SlavsyaRus, Slightsmile, Slon02, SonicAD, Sorrat, Sotnik, Space Cadet, Sparthorse, Spitfire, Squiddy, Standard888, SteinbDJ, Supermacey, Supernaturalist, Surfingslovak, Taamu, Takeshi0928, Tanaats, TarzanASG, Tasc, Taton80, Tbhotch, Tcncv, Te5, Techman224, Technical 13, Teevek, Telempe, Telex, Telfordbuck, Template namespace initialisation script, Tewfik, The Evil Spartan, Thecalogero0120, Thehelpfulone, Therexbanner, Thingg, Tide rolls, TimBentley, Tizio, Tobby72, Tombseye, Tomcat7, Tommy2010, Tone, TotalSpaceshipGuy3, Trasman, Travellingnorth, Troltroltrol, Trueshow111, Trusilver, Tt 225, Turkishbrat, Tycuddy, VIP2009, VKokielov, VNCCC, Valip, Vals gay, Vaniba12, Vanished user aqpoi4u3tijsrfi, Varanwal, Velázquez, Vianello, Vitaxa o8, Vlarus yarilo, Vmrgrsergr, Voevoda, Volykr yilevas, Vornoff, Voyevoda, Vufimtsev, Vviralpapillome, WadeSimMiser, Wadim, Wayne Slam, Werdna, WereSpielChequers, West.andrew.g, Wester, WhisperToMe, Wifione, Wiki alf, Wikisanchez, Wing, Wingbac22, Woohookitty, Wtmitchell, Xaxosax, Xi HACK3R ix, Yair rand, Ybelov, Yeltsinfan, YeshuaDavid, Yulia Romero, Zaparojdik, Zenit4ever, Zlerman, Zndawg, Zocky, Zohebj, Zpird, Zundark, Zyma, Губин Михаил, Куниса, ОйЛ, Хинт, Шиманський Василь, ینوگنابنوگن, 1258 anonymous edits

Time_in_Russia *Source*: http://en.wikipedia.org/w/index.php?title=Time_in_Russia *Contributors*: A.h. king, AlanM1, Alansohn, BIL, Bonzon, Bridgetansel, Chrism, CommonsDelinker, Coolhawks88, Coroboy, D.M. from Ukraine, Devilhope, Dewritech, Double sharp, Ekren, Eldumpo, Eubulides, Ezhiki, Geanixx, Gestumblindi, Ghirlandajo, Goudzovski, Greyhood, HTML2011, Hafrul, Hardscarf, Haywire, Hellerick, Heracletus, Incnis Mrsi, Indefatigable, January2007, JohnCD, JohnnyWiki, Kahastok, Khazar, Klemen Kocjancic, Kmorozov, Kostja, Kwamikagami, Matthew hk, Maximaximax, Mikemoral, Neelix, Orange-kun, Physchim62, Prisonermonkeys, PsyMar, R9tgokunks, RadioKirk, RedYeti, Reywas92, Rklawton, Roentgenium111, Royaume du Maroc, TZ master, TarzanASG, Therexbanner, Tim Parenti, Time in Russia, TimeCodex, TimeCurrency, TimeXCode, Tobias Conradi, TopoChecker, Trust Is All You Need, Ultimate Destiny, Woohookitty, Wxidea, ZanderSchubert, Zzyzx11, 38 anonymous edits

Image Sources, Licenses and Contributors

File:Flag of Perm Krai.svg *Source*: http://en.wikipedia.org/w/index.php?title=File:Flag_of_Perm_Krai.svg *License*: unknown *Contributors*: User:const_st

File:Coat of Arms of Perm Krai.svg *Source*: http://en.wikipedia.org/w/index.php?title=File:Coat_of_Arms_of_Perm_Krai.svg *License*: unknown *Contributors*: Margarita Tarasova; based on historical coat of arms, adopted in 1783

File:Perm in Russia.svg *Source*: http://en.wikipedia.org/w/index.php?title=File:Perm_in_Russia.svg *License*: unknown *Contributors*: TUBS

File:Maksimovsky rock Chusovaya river.jpg *Source*: http://en.wikipedia.org/w/index.php?title=File:Maksimovsky_rock_Chusovaya_river.jpg *License*: unknown *Contributors*: Sergei Mikhailovich Prokudin-Gorskii

File:Vishera.jpg *Source*: http://en.wikipedia.org/w/index.php?title=File:Vishera.jpg *License*: unknown *Contributors*: Haneburger, Hugo.arg, Pline, Putnik

File:Salt dump in Solikamsk.jpg *Source*: http://en.wikipedia.org/w/index.php?title=File:Salt_dump_in_Solikamsk.jpg *License*: unknown *Contributors*: User:Мастер Снов

File:Cave Kungurskaya 0091.JPG *Source*: http://en.wikipedia.org/w/index.php?title=File:Cave_Kungurskaya_0091.JPG *License*: unknown *Contributors*: User:Nzeemin

File:Perm Krai Legislative Assembly.jpg *Source*: http://en.wikipedia.org/w/index.php?title=File:Perm_Krai_Legislative_Assembly.jpg *License*: unknown *Contributors*: User:Bestalex

File:Въезд в Чернушинский район.jpg *Source*: http://en.wikipedia.org/w/index.php?title=File:Въезд_в_Чернушинский_район.jpg *License*: unknown *Contributors*: User:AndreyIGOSHEV

File:Kamskaya Hydro Power Station.jpg *Source*: http://en.wikipedia.org/w/index.php?title=File:Kamskaya_Hydro_Power_Station.jpg *License*: unknown *Contributors*: User:Мастер Снов

File:Perm museum.jpg *Source*: http://en.wikipedia.org/w/index.php?title=File:Perm_museum.jpg *License*: unknown *Contributors*: shakko

File:Khokhlovka.jpg *Source*: http://en.wikipedia.org/w/index.php?title=File:Khokhlovka.jpg *License*: unknown *Contributors*: User:Мастер Снов

File:Flag of Russia.svg *Source*: http://en.wikipedia.org/w/index.php?title=File:Flag_of_Russia.svg *License*: unknown *Contributors*: Anomie, Zscout370

File:Coat of Arms of the Russian Federation.svg *Source*: http://en.wikipedia.org/w/index.php?title=File:Coat_of_Arms_of_the_Russian_Federation.svg *License*: unknown *Contributors*: Федеральный конституционный закон «О Государственном гербе Российской Федерации»

File:Russian Federation (orthographic projection).svg *Source*: http://en.wikipedia.org/w/index.php?title=File:Russian_Federation_(orthographic_projection).svg *License*: unknown *Contributors*: User:Ssolbergj

File:Increase2.svg *Source*: http://en.wikipedia.org/w/index.php?title=File:Increase2.svg *License*: unknown *Contributors*: Sarang

file:speakerlink-new.svg *Source*: http://en.wikipedia.org/w/index.php?title=File:Speakerlink-new.svg *License*: unknown *Contributors*: User:Kelvinsong

File:Speaker Icon.svg *Source*: http://en.wikipedia.org/w/index.php?title=File:Speaker_Icon.svg *License*: unknown *Contributors*: Blast, G.Hagedorn, Mobius, Tehdog, 2 anonymous edits

File:IE expansion.png *Source*: http://en.wikipedia.org/w/index.php?title=File:IE_expansion.png *License*: unknown *Contributors*: Dbachmann, Esteban.barahona, Man vyi, Mdd, Mvangeest, Ricordisamoa, 1 anonymous edits

File:Kievan Rus en.jpg *Source*: http://en.wikipedia.org/w/index.php?title=File:Kievan_Rus_en.jpg *License*: unknown *Contributors*: Captain Blood, Kintetsubuffalo, Ras67, Rottweiler, Silar

File:Lebedev baptism.jpg *Source*: http://en.wikipedia.org/w/index.php?title=File:Lebedev_baptism.jpg *License*: unknown *Contributors*: Bogomolov.PL, Butko, Joseolgon, Mladifilozof, Randbewohner, Vizu, 1 anonymous edits

File:Lissner TroiceSergievaLavr.jpg *Source*: http://en.wikipedia.org/w/index.php?title=File:Lissner_TroiceSergievaLavr.jpg *License*: unknown *Contributors*: Alex Bakharev, Art-top, Dmitry Rozhkov, MishaPan, Pianist, Shakko, 1 anonymous edits

File:Ivan the Terrible (cropped).JPG *Source*: http://en.wikipedia.org/w/index.php?title=File:Ivan_the_Terrible_(cropped).JPG *License*: unknown *Contributors*: Bukk, Butko, Kürschner, Mrlopez2681, Shakko

File:Minin&Pogjarsky 2.jpg *Source*: http://en.wikipedia.org/w/index.php?title=File:Minin&Pogjarsky_2.jpg *License*: unknown *Contributors*: EugeneZelenko, Julmin, Kneiphof, Okman, 1 anonymous edits

File:Peter der-Grosse 1838.jpg *Source*: http://en.wikipedia.org/w/index.php?title=File:Peter_der-Grosse_1838.jpg *License*: unknown *Contributors*: Anathema, Butko, Darwinius, Ecummenic, Kilom691, Mathiasrex, Mattes, Miyokan, Polarlys, Sammyday, Shakko, Sparkit, TheoClarke, Vincent Steenberg, Zoloto-Russland

File:Russian Empire (orthographic projection).svg *Source*: http://en.wikipedia.org/w/index.php?title=File:Russian_Empire_(orthographic_projection).svg *License*: unknown *Contributors*: User:Shadowxfox

File:Lenintribune.jpg *Source*: http://en.wikipedia.org/w/index.php?title=File:Lenintribune.jpg *License*: unknown *Contributors*: Original uploader was Black31 at fr.wikipedia

File:Russia-2000-stamp-Tatlin Tower and Worker and Kolkhoz Woman by Vera Mukhina.jpg *Source*: http://en.wikipedia.org/w/index.php?title=File:Russia-2000-stamp-Tatlin_Tower_and_Worker_and_Kolkhoz_Woman_by_Vera_Mukhina.jpg *License*: unknown *Contributors*: Butko, Editor at Large, Gryffindor, Man vyi, Michael Romanov, Philafrenzy, Shakko, Zimin.V.G.

File:Red Army greeted in Bucharest.jpg *Source*: http://en.wikipedia.org/w/index.php?title=File:Red_Army_greeted_in_Bucharest.jpg *License*: unknown *Contributors*: Bogdan, CaptainFugu, Dahn, Jmabel, John commons, Teofilo, 2 anonymous edits

File:Convert ru kosmos077.jpg *Source*: http://en.wikipedia.org/w/index.php?title=File:Convert_ru_kosmos077.jpg *License*: unknown *Contributors*: USSR government

File:Sputnik asm.jpg *Source*: http://en.wikipedia.org/w/index.php?title=File:Sputnik_asm.jpg *License*: unknown *Contributors*: User:Jappalang

File:Moscow, City May 2010 03.JPG *Source*: http://en.wikipedia.org/w/index.php?title=File:Moscow,_City_May_2010_03.JPG *License*: unknown *Contributors*: User:NVO

File:Moscow Kremlin from Kamenny bridge.jpg *Source*: http://en.wikipedia.org/w/index.php?title=File:Moscow_Kremlin_from_Kamenny_bridge.jpg *License*: unknown *Contributors*: User:Julmin, User:Surendil

File:BRIC leaders in 2008.jpg *Source*: http://en.wikipedia.org/w/index.php?title=File:BRIC_leaders_in_2008.jpg *License*: unknown *Contributors*: Presidential Press and Information Office

File:Countries-SCO.png *Source*: http://en.wikipedia.org/w/index.php?title=File:Countries-SCO.png *License*: unknown *Contributors*: User:Sejogo

File:2010 Moscow Victory Day Parade-5.jpeg *Source*: http://en.wikipedia.org/w/index.php?title=File:2010_Moscow_Victory_Day_Parade-5.jpeg *License*: unknown *Contributors*: Presidential Press and Information Office

File:Russian Regions-EN.svg *Source*: http://en.wikipedia.org/w/index.php?title=File:Russian_Regions-EN.svg *License*: unknown *Contributors*: User:Ezhiki, User:Plavius, User:Tavork

File:Russland Relief.png *Source*: http://en.wikipedia.org/w/index.php?title=File:Russland_Relief.png *License*: unknown *Contributors*: User:Lencer

File:Elbrus North 195.jpg *Source*: http://en.wikipedia.org/w/index.php?title=File:Elbrus_North_195.jpg *License*: unknown *Contributors*: Dmitrij96, Herzi Pinki, Iltever, Kykovner, 1 anonymous edits

File:Осетр река.jpg *Source*: http://en.wikipedia.org/w/index.php?title=File:Осетр_река.jpg *License*: unknown *Contributors*: Hugo.arg, Kolossus, Lvova

File:Archangelsk taiga.JPG *Source*: http://en.wikipedia.org/w/index.php?title=File:Archangelsk_taiga.JPG *License*: unknown *Contributors*: User:Xxxl

File:Medved mzoo.jpg *Source*: http://en.wikipedia.org/w/index.php?title=File:Medved_mzoo.jpg *License*: unknown *Contributors*: User:Simm

File:WTC-Moscow.jpg *Source*: http://en.wikipedia.org/w/index.php?title=File:WTC-Moscow.jpg *License*: unknown *Contributors*: User:Wecameasromans

File:Russian economy since fall of Soviet Union.PNG *Source*: http://en.wikipedia.org/w/index.php?title=File:Russian_economy_since_fall_of_Soviet_Union.PNG *License*: unknown *Contributors*: User:LokiiT

File:Rozh.jpg *Source*: http://en.wikipedia.org/w/index.php?title=File:Rozh.jpg *License*: unknown *Contributors*: Afanasovich, Ain92, Bdk, Boo-Boo Baroo, Ghirlandajo, Ixtzib, Julia W, Le.Loup.Gris, Neukoln, Sfrandzi, Shakko, Vhorvat

File:RF NG pipestoEU.gif *Source*: http://en.wikipedia.org/w/index.php?title=File:RF_NG_pipestoEU.gif *License*: unknown *Contributors*: DOE

File:TransSiberianRailwayAtKm9288.jpg *Source*: http://en.wikipedia.org/w/index.php?title=File:TransSiberianRailwayAtKm9288.jpg *License*: unknown *Contributors*: Bukk, Bulwersator, Demidow, Ingolfson, JuTa, Krinkle, Mschlindwein, Ymblanter, 1 anonymous edits

File:NSF picture of Yamal.gif *Source*: http://en.wikipedia.org/w/index.php?title=File:NSF_picture_of_Yamal.gif *License*: unknown *Contributors*: High Contrast, Kneiphof, Maximaximax, Stunteltje, Tungsten, 1 anonymous edits

File:Lomonosovportrait.jpg *Source*: http://en.wikipedia.org/w/index.php?title=File:Lomonosovportrait.jpg *License*: unknown *Contributors*: Anne97432, Ecummenic, Hsarrazin, Jflomonosoff, Papa November, Shakko

File:Soyuz TMA-2 launch.jpg *Source*: http://en.wikipedia.org/w/index.php?title=File:Soyuz_TMA-2_launch.jpg *License*: unknown *Contributors*: NASA/Scott Andrews

File:Sukhoi T-50 Beltyukov.jpg *Source*: http://en.wikipedia.org/w/index.php?title=File:Sukhoi_T-50_Beltyukov.jpg *License*: unknown *Contributors*: Alex Beltyukov

File:Yo-conc3.JPG *Source*: http://en.wikipedia.org/w/index.php?title=File:Yo-conc3.JPG *License*: unknown *Contributors*: User:Brattarb

File:Mir on 12 June 1998edit1.jpg *Source*: http://en.wikipedia.org/w/index.php?title=File:Mir_on_12_June_1998edit1.jpg *License*: unknown *Contributors*: User:Thegreenj

Image:Percentage of Russians by region.svg *Source*: http://en.wikipedia.org/w/index.php?title=File:Percentage_of_Russians_by_region.svg *License*: unknown *Contributors*: User:Любослов Езыкин

Image:Russia natural population growth rates 2012.PNG *Source*: http://en.wikipedia.org/w/index.php?title=File:Russia_natural_population_growth_rates_2012.PNG *License*: unknown *Contributors*: User:Anvyder

File:RussianLanguageMap.png *Source*: http://en.wikipedia.org/w/index.php?title=File:RussianLanguageMap.png *License*: unknown *Contributors*: BartekChom, Chinneeb, Damian Radu, David Kernow, DenghiùComm, JuTa, Kuban kazak, Maphobbyist, NuclearVacuum, Roke, Small Bug, Spesh531, Tat1642, Wester, Winterheart, 22 anonymous edits

File:Kazan church edit1.jpg *Source*: http://en.wikipedia.org/w/index.php?title=File:Kazan_church_edit1.jpg *License*: unknown *Contributors*: Maarten

File:Terapevt Mudrov train.JPG *Source*: http://en.wikipedia.org/w/index.php?title=File:Terapevt_Mudrov_train.JPG *License*: unknown *Contributors*: User:Glucke

File:Школа 1118 (Москва).jpg *Source*: http://en.wikipedia.org/w/index.php?title=File:Школа_1118_(Москва).jpg *License*: unknown *Contributors*: vow

File:Kustodiev Merchants Wife.jpg *Source*: http://en.wikipedia.org/w/index.php?title=File:Kustodiev_Merchants_Wife.jpg *License*: unknown *Contributors*: Alex Bakharev, AndreasPraefcke, Bukk, Jonkerz, Jtir, Kaganer, Mattbuck, Mattes, Moogsi, Pierpao, Pitke, Shakko, SunOfErat, Talmoryair, Thorvaldsson, Vincent Steenberg, Wst

File:RedSquare SaintBasile (pixinn.net).jpg *Source*: http://en.wikipedia.org/w/index.php?title=File:RedSquare_SaintBasile_(pixinn.net).jpg *License*: unknown *Contributors*: User:XtoF

File:Andrej Rublëv 001.jpg *Source*: http://en.wikipedia.org/w/index.php?title=File:Andrej_Rublëv_001.jpg *License*: unknown *Contributors*: AndreasPraefcke, AnonMoos, Aroche, Goldfritha, Laurom, Mattes, Shakko, Tsca, Wst

File:Snowdance.jpg *Source*: http://en.wikipedia.org/w/index.php?title=File:Snowdance.jpg *License*: unknown *Contributors*: Rick Dikeman

File:L.N.Tolstoy Prokudin-Gorsky.jpg *Source*: http://en.wikipedia.org/w/index.php?title=File:L.N.Tolstoy_Prokudin-Gorsky.jpg *License*: unknown *Contributors*: Avenger911, Berillium, Ctac, Damouns, Ecummenic, GreatOrangePumpkin, Henry Merrivale, J.M.Domingo, Methem, MrPanyGoff, Oldcolor, Pianist, RedAndr, Reino Həlismaa, Sasha Krotov, Tommy Kronkvist, Wolfmann, Yann, 3 anonymous edits

File:Medvedev - Russia Today 3.jpg *Source*: http://en.wikipedia.org/w/index.php?title=File:Medvedev_-_Russia_Today_3.jpg *License*: unknown *Contributors*: Presidential Press and Information Office

File:Shukhov tower shabolovka moscow 02.jpg *Source*: http://en.wikipedia.org/w/index.php?title=File:Shukhov_tower_shabolovka_moscow_02.jpg *License*: unknown *Contributors*: Arssenev, Basilicofresco, Donskoy, Lite, NVO, Rlevse, Roomba, Snowgrove, TomAlt, Vonvon, 5 anonymous edits

File:RussianBear.JPG *Source*: http://en.wikipedia.org/w/index.php?title=File:RussianBear.JPG *License*: unknown *Contributors*: Roman Kovrigin

File:Dmitry Medvedev 20 May 2008-2.jpg *Source*: http://en.wikipedia.org/w/index.php?title=File:Dmitry_Medvedev_20_May_2008-2.jpg *License*: unknown *Contributors*: Presidential Press and Information Office

File:Victory Day Parade 2005-18.jpg *Source*: http://en.wikipedia.org/w/index.php?title=File:Victory_Day_Parade_2005-18.jpg *License*: unknown *Contributors*: ITAR-TASS

File:Alyye parusa.jpg *Source*: http://en.wikipedia.org/w/index.php?title=File:Alyye_parusa.jpg *License*: unknown *Contributors*: Marina Lystseva

File:PeterhofGrandCascade.JPG *Source*: http://en.wikipedia.org/w/index.php?title=File:PeterhofGrandCascade.JPG *License*: unknown *Contributors*: Cottbus, Kaganer, Ronaldino

File:Russian-Matroshka2.jpg *Source*: http://en.wikipedia.org/w/index.php?title=File:Russian-Matroshka2.jpg *License*: unknown *Contributors*: Fanghong, Ingolfson, Mariluna

File:Gnome-globe.svg *Source*: http://en.wikipedia.org/w/index.php?title=File:Gnome-globe.svg *License*: unknown *Contributors*: David Vignoni

File:Map of Russian districts, 2010-01-19.svg *Source*: http://en.wikipedia.org/w/index.php?title=File:Map_of_Russian_districts,_2010-01-19.svg *License*: unknown *Contributors*: User:SeNeKa, user:Lokal_Profil

File:Economic regions of Russia.png *Source*: http://en.wikipedia.org/w/index.php?title=File:Economic_regions_of_Russia.png *License*: unknown *Contributors*: User:Nightstallion

File:COB data Russia.PNG *Source*: http://en.wikipedia.org/w/index.php?title=File:COB_data_Russia.PNG *License*: unknown *Contributors*: User:Kransky

File:Russian census 2010 - stationary center.jpg *Source*: http://en.wikipedia.org/w/index.php?title=File:Russian_census_2010_-_stationary_center.jpg *License*: unknown *Contributors*: User:Schekinov Alexey Victorovich

File:RR5514-0072R.png *Source*: http://en.wikipedia.org/w/index.php?title=File:RR5514-0072R.png *License*: unknown *Contributors*: EvilTeeth, Викимонетчик

File:Chuvash diaspora in Volga Federal District.png *Source*: http://en.wikipedia.org/w/index.php?title=File:Chuvash_diaspora_in_Volga_Federal_District.png *License*: unknown *Contributors*: User:PCode

File:Russian_Winter_Festival_London_2007_121.jpg *Source*: http://en.wikipedia.org/w/index.php?title=File:Russian_Winter_Festival_London_2007_121.jpg *License*: unknown *Contributors*: S Pakhrin

File:Russian Winter Festival London 2007 119.jpg *Source*: http://en.wikipedia.org/w/index.php?title=File:Russian_Winter_Festival_London_2007_119.jpg *License*: unknown *Contributors*: S Pakhrin

File:Чувашский_праздник1.jpg *Source*: http://en.wikipedia.org/w/index.php?title=File:Чувашский_праздник1.jpg *License*: unknown *Contributors*: User:Safary07

File:Средненизовые_чуваши_"анат_енчи".jpeg *Source*: http://en.wikipedia.org/w/index.php?title=File:Средненизовые_чуваши_"анат_енчи".jpeg *License*: unknown *Contributors*: User:Safary07

File:Тухья анат енчи.jpg *Source*: http://en.wikipedia.org/w/index.php?title=File:Тухья_анат_енчи.jpg *License*: unknown *Contributors*: User:Safary07

File:Gorskii 04421u.jpg *Source*: http://en.wikipedia.org/w/index.php?title=File:Gorskii_04421u.jpg *License*: unknown *Contributors*: , digital rendering for the Library of Congress by Walter Frankhauser / WalterStudio

file:Russophone.svg *Source*: http://en.wikipedia.org/w/index.php?title=File:Russophone.svg *License*: unknown *Contributors*: User:NuclearVacuum

File:Eglise notre dame de l assomption 7.jpg *Source*: http://en.wikipedia.org/w/index.php?title=File:Eglise_notre_dame_de_l_assomption_7.jpg *License*: unknown *Contributors*: Olivier Perrin Original uploader was O.perrin at fr.wikipedia

File:Russian Orthodox Church, Old Shanghai.jpg *Source*: http://en.wikipedia.org/w/index.php?title=File:Russian_Orthodox_Church,_Old_Shanghai.jpg *License*: unknown *Contributors*: stevechasmar

File:Russians ethnic 94.jpg *Source*: http://en.wikipedia.org/w/index.php?title=File:Russians_ethnic_94.jpg *License*: unknown *Contributors*: Altenmann, High Contrast, Nachcommonsverschieber, Sevela.p, Sibom

File:Gagarin in Sweden.jpg *Source*: http://en.wikipedia.org/w/index.php?title=File:Gagarin_in_Sweden.jpg *License*: unknown *Contributors*: Arkiv: Sydsvenskan

File:UTC hue4map RUS 2011-09-01.svg *Source*: http://en.wikipedia.org/w/index.php?title=File:UTC_hue4map_RUS_2011-09-01.svg *License*: unknown *Contributors*: User:Lokal Profil, User:TZ master

File:UTC hue4map RUS 2009.svg *Source*: http://en.wikipedia.org/w/index.php?title=File:UTC_hue4map_RUS_2009.svg *License*: unknown *Contributors*: User:TZ master, user:Lokal_Profil

File:UTC hue4map RUS 2010-04-01.svg *Source*: http://en.wikipedia.org/w/index.php?title=File:UTC_hue4map_RUS_2010-04-01.svg *License*: unknown *Contributors*: User:TZ master, user:Lokal_Profil

File:UTC hue4map RUS-SA 2010.svg *Source*: http://en.wikipedia.org/w/index.php?title=File:UTC_hue4map_RUS-SA_2010.svg *License*: unknown *Contributors*: User:Lokal_Profil, User:TZ master

File:UTC hue4map RUS-SA 2011-09-01.svg *Source*: http://en.wikipedia.org/w/index.php?title=File:UTC_hue4map_RUS-SA_2011-09-01.svg *License*: unknown *Contributors*: User:Lokal_Profil, User:TZ master

File:Tz map russia2009r efeledotnet.png *Source*: http://en.wikipedia.org/w/index.php?title=File:Tz_map_russia2009r_efeledotnet.png *License*: unknown *Contributors*: Eric Muller

. PREAMBLE

The purpose of this License is to make a manual, textbook, or ›ther functional and useful document "free" in the sense of ·eedom: to assure everyone the effective freedom to copy and ›edistribute it, with or without modifying it, either commercially or ›oncommercially. Secondarily, this License preserves for the ›uthor and publisher a way to get credit for their work, while not ›eing considered responsible for modifications made by others. ʼhis License is a kind of "copyleft", which means that derivative ›vorks of the document must themselves be free in the same ›ense. It complements the GNU General Public License, which is › copyleft license designed for free software. We have designed ›his License in order to use it for manuals for free software, ›ecause free software needs free documentation: a free program ›hould come with manuals providing the same freedoms that the ›oftware does. But this License is not limited to software manuals; : can be used for any textual work, regardless of subject matter ›r whether it is published as a printed book. We recommend this ›icense principally for works whose purpose is instruction or ›eference.

. APPLICABILITY AND DEFINITIONS

This License applies to any manual or other work, in any ›nedium, that contains a notice placed by the copyright holder ›aying it can be distributed under the terms of this License. Such › notice grants a world-wide, royalty-free license, unlimited in ›uration, to use that work under the conditions stated herein. The Document", below, refers to any such manual or work. Any ›nember of the public is a licensee, and is addressed as "you". ʼou accept the license if you copy, modify or distribute the work ›n a way requiring permission under copyright law. A "Modified ›ersion" of the Document means any work containing the Document or a portion of it, either copied verbatim, or with ›nodifications and/or translated into another language. A Secondary Section" is a named appendix or a front-matter ›ection of the Document that deals exclusively with the ›elationship of the publishers or authors of the Document to the Document's overall subject (or to related matters) and contains ›othing that could fall directly within that overall subject. (Thus, if ›he Document is in part a textbook of mathematics, a Secondary Section may not explain any mathematics.) The relationship could ›e a matter of historical connection with the subject or with ›elated matters, or of legal, commercial, philosophical, ethical or ›olitical position regarding them. The "Invariant Sections" are ›ertain Secondary Sections whose titles are designated, as being those of Invariant Sections, in the notice that says that the Document is released under this License. If a section does not fit the above definition of Secondary then it is not allowed to be designated as Invariant. The Document may contain zero Invariant Sections. If the Document does not identify any Invariant Sections then there are none. The "Cover Texts" are certain short passages of text that are listed, as Front-Cover Texts or Back-Cover Texts, in the notice that says that the Document is released under this License. A Front-Cover Text may be at most 5 words, and a Back-Cover Text may be at most 25 words. A "Transparent" copy of the Document means a machine-readable copy, represented in a format whose specification is available to the general public, that is suitable for revising the document straightforwardly with generic text editors or (for images composed of pixels) generic paint programs or (for drawings) some widely available drawing editor, and that is suitable for input to text formatters or for automatic translation to a variety of formats suitable for input to text formatters. A copy made in an otherwise Transparent file format whose markup, or absence of markup, has been arranged to thwart or discourage subsequent modification by readers is not Transparent. An image format is not Transparent if used for any substantial amount of text. A copy that is not "Transparent" is called "Opaque". Examples of suitable formats for Transparent copies include plain ASCII without markup, Texinfo input format, LaTeX input format, SGML or XML using a publicly available DTD, and standard-conforming simple HTML, PostScript or PDF designed for human modification. Examples of transparent image formats include PNG, XCF and JPG. Opaque formats include proprietary formats that can be read and edited only by proprietary word processors, SGML or XML for which the DTD and/or processing tools are not generally available, and the machine-generated HTML, PostScript or PDF produced by some word processors for output purposes only. The "Title Page" means, for a printed book, the title page itself, plus such following pages as are needed to hold, legibly, the material this License requires to appear in the title page. For works in formats which do not have any title page as such, "Title Page" means the text near the most prominent appearance of the work's title, preceding the beginning of the body of the text. A section "Entitled XYZ" means a named subunit of the Document whose title either is precisely XYZ or contains XYZ in parentheses following text that translates XYZ in another language. (Here XYZ stands for a specific section name mentioned below, such as "Acknowledgements", "Dedications", "Endorsements", or "History".) To "Preserve the Title" of such a section when you modify the Document means that it remains a section "Entitled XYZ" according to this definition. The Document may include Warranty Disclaimers next to the notice which states that this License applies to the Document. These Warranty Disclaimers are considered to be included by reference in this License, but only as regards disclaiming warranties: any other implication that these Warranty Disclaimers may have is void and has no effect on the meaning of this License.

2. VERBATIM COPYING

You may copy and distribute the Document in any medium, either commercially or noncommercially, provided that this License, the copyright notices, and the license notice saying this License applies to the Document are reproduced in all copies, and that you add no other conditions whatsoever to those of this License. You may not use technical measures to obstruct or control the reading or further copying of the copies you make or distribute. However, you may accept compensation in exchange for copies. If you distribute a large enough number of copies you must also follow the conditions in section 3. You may also lend copies, under the same conditions stated above, and you may publicly display copies.

3. COPYING IN QUANTITY

If you publish printed copies (or copies in media that commonly have printed covers) of the Document, numbering more than 100, and the Document's license notice requires Cover Texts, you must enclose the copies in covers that carry, clearly and legibly, all these Cover Texts: Front-Cover Texts on the front cover, and Back-Cover Texts on the back cover. Both covers must also clearly and legibly identify you as the publisher of these copies. The front cover must present the full title with all words of the title equally prominent and visible. You may add other material on the covers in addition. Copying with changes limited to the covers, as long as they preserve the title of the Document and satisfy these conditions, can be treated as verbatim copying in other respects. If the required texts for either cover are too voluminous to fit legibly, you should put the first ones listed (as many as fit reasonably) on the actual cover, and continue the rest onto adjacent pages. If you publish or distribute Opaque copies of the Document numbering more than 100, you must either include a machine-readable Transparent copy along with each Opaque copy, or state in or with each Opaque copy a computer-network location from which the general network-using public has access to download using public-standard network protocols a complete Transparent copy of the Document, free of added material. If you use the latter option, you must take reasonably prudent steps, when you begin distribution of Opaque copies in quantity, to ensure that this Transparent copy will remain thus accessible at the stated location until at least one year after the last time you distribute an Opaque copy (directly or through your agents or retailers) of that edition to the public. It is requested, but not required, that you contact the authors of the Document well before redistributing any large number of copies, to give them a chance to provide you with an updated version of the Document.

4. MODIFICATIONS

You may copy and distribute a Modified Version of the Document under the conditions of sections 2 and 3 above, provided that you release the Modified Version under precisely this License, with the Modified Version filling the role of the Document, thus licensing distribution and modification of the Modified Version to whoever possesses a copy of it. In addition, you must do these things in the Modified Version: A. Use in the Title Page (and on the covers, if any) a title distinct from that of the Document, and from those of previous versions (which should, if there were any, be listed in the History section of the Document). You may use the same title as a previous version if the original publisher of that version gives permission. B. List on the Title Page, as authors, one or more persons or entities responsible for authorship of the modifications in the Modified Version, together with at least five of the principal authors of the Document (all of its principal authors, if it has fewer than five), unless they release you from this requirement. C. State on the Title page the name of the publisher of the Modified Version, as the publisher. D. Preserve all the copyright notices of the Document. E. Add an appropriate copyright notice for your modifications adjacent to the other copyright notices. F. Include, immediately after the copyright notices, a license notice giving the public permission to use the Modified Version under the terms of this License, in the form shown in the Addendum below. G. Preserve in that license notice the full lists of Invariant Sections and required Cover Texts given in the Document's license notice. H. Include an unaltered copy of this License. I. Preserve the section Entitled "History", Preserve its Title, and add to it an item stating at least the title, year, new authors, and publisher of the Modified Version as given on the Title Page. If there is no section Entitled "History" in the Document, create one stating the title, year, authors, and publisher of the Document as given on its Title Page, then add an item describing the Modified Version as stated in the previous sentence. J. Preserve the network location, if any, given in the Document for public access to a Transparent copy of the Document, and likewise the network locations given in the Document for previous versions it was based on. These may be placed in the "History" section. You may omit a network location for a work that was published at least four years before the Document itself, or if the original publisher of the version it refers to gives permission. K. For any section Entitled "Acknowledgements" or "Dedications", Preserve the Title of the section, and preserve in the section all the substance and tone of each of the contributor acknowledgements and/or dedications given therein. L. Preserve all the Invariant Sections of the Document, unaltered in their text and in their titles. Section numbers or the equivalent are not considered part of the section titles. M. Delete any section Entitled "Endorsements". Such a section may not be included in the Modified Version. N. Do not retitle any existing section to be Entitled "Endorsements" or to conflict in title with any Invariant Section. O. Preserve any Warranty Disclaimers. If the Modified Version includes new front-matter sections or appendices that qualify as Secondary Sections and contain no material copied from the Document, you may at your option designate some or all of these sections as invariant. To do this, add their titles to the list of Invariant Sections in the Modified Version's license notice. These titles must be distinct from any other section titles. You may add a section Entitled "Endorsements", provided it contains nothing but endorsements of your Modified Version by various parties--for example, statements of peer review or that the text has been approved by an organization as the authoritative definition of a standard. You may add a passage of up to five words as a Front-Cover Text, and a passage of up to 25 words as a Back-Cover Text, to the end of the list of Cover Texts in the Modified Version. Only one passage of Front-Cover Text and one of Back-Cover Text may be added by (or through arrangements made by) any one entity. If the Document already includes a cover text for the same cover, previously added by you or by arrangement made by the same entity you are acting on behalf of, you may not add another; but you may replace the old one, on explicit permission from the previous publisher that added the old one. The author(s) and publisher(s) of the Document do not by this License give permission to use their names for publicity for or to assert or imply endorsement of any Modified Version.

5. COMBINING DOCUMENTS

You may combine the Document with other documents released under this License, under the terms defined in section 4 above for modified versions, provided that you include in the combination all of the Invariant Sections of all of the original documents, unmodified, and list them all as Invariant Sections of your combined work in its license notice, and that you preserve all their Warranty Disclaimers. The combined work need only contain one copy of this License, and multiple identical Invariant Sections may be replaced with a single copy. If there are multiple Invariant Sections with the same name but different contents, make the title of each such section unique by adding at the end of it, in parentheses, the name of the original author or publisher of that section if known, or else a unique number. Make the same adjustment to the section titles in the list of Invariant Sections in the license notice of the combined work. In the combination, you must combine any sections Entitled "History" in the various original documents, forming one section Entitled "History"; likewise combine any sections Entitled "Acknowledgements", and any sections Entitled "Dedications". You must delete all sections Entitled "Endorsements".

6. COLLECTIONS OF DOCUMENTS

You may make a collection consisting of the Document and other documents released under this License, and replace the individual copies of this License in the various documents with a single copy that is included in the collection, provided that you follow the rules of this License for verbatim copying of each of the documents in all other respects. You may extract a single document from such a collection, and distribute it individually under this License, provided you insert a copy of this License into the extracted document, and follow this License in all other respects regarding verbatim copying of that document.

7. AGGREGATION WITH INDEPENDENT WORKS

A compilation of the Document or its derivatives with other separate and independent documents or works, in or on a volume of a storage or distribution medium, is called an "aggregate" if the copyright resulting from the compilation is not used to limit the legal rights of the compilation's users beyond what the individual works permit. When the Document is included in an aggregate, this License does not apply to the other works in the aggregate which are not themselves derivative works of the Document. If the Cover Text requirement of section 3 is applicable to these copies of the Document, then if the Document is less than one half of the entire aggregate, the Document's Cover Texts may be placed on covers that bracket the Document within the aggregate, or the electronic equivalent of covers if the Document is in electronic form. Otherwise they must appear on printed covers that bracket the whole aggregate.

8. TRANSLATION

Translation is considered a kind of modification, so you may distribute translations of the Document under the terms of section 4. Replacing Invariant Sections with translations requires special permission from their copyright holders, but you may include translations of some or all Invariant Sections in addition to the original versions of these Invariant Sections. You may include a translation of this License, and all the license notices in the Document, and any Warranty Disclaimers, provided that you also include the original English version of this License and the original versions of those notices and disclaimers. In case of a disagreement between the translation and the original version of this License or a notice or disclaimer, the original version will prevail. If a section in the Document is Entitled "Acknowledgements", "Dedications", or "History", the requirement (section 4) to Preserve its Title (section 1) will typically require changing the actual title.

9. TERMINATION

You may not copy, modify, sublicense, or distribute the Document except as expressly provided for under this License. Any other attempt to copy, modify, sublicense or distribute the Document is void, and will automatically terminate your rights under this License. However, parties who have received copies, or rights, from you under this License will not have their licenses terminated so long as such parties remain in full compliance.

10. FUTURE REVISIONS OF THIS LICENSE

The Free Software Foundation may publish new, revised versions of the GNU Free Documentation License from time to time. Such new versions will be similar in spirit to the present version, but may differ in detail to address new problems or concerns. See http://www.gnu.org/copyleft/. Each version of the License is given a distinguishing version number. If the Document specifies that a particular numbered version of this License "or any later version" applies to it, you have the option of following the terms and conditions either of that specified version or of any later version that has been published (not as a draft) by the Free Software Foundation. If the Document does not specify a version number of this License, you may choose any version ever published (not as a draft) by the Free Software Foundation. ADDENDUM: How to use this License for your documents To use this License in a document you have written, include a copy of the License in the document and put the following copyright and license notices just after the title page: Copyright (c) YEAR YOUR NAME. Permission is granted to copy, distribute and/or modify this document under the terms of the GNU Free Documentation License, Version 1.2 or any later version published by the Free Software Foundation; with no Invariant Sections, no Front-Cover Texts, and no Back-Cover Texts. A copy of the license is included in the section entitled "GNU Free Documentation License". If you have Invariant Sections, Front-Cover Texts and Back-Cover Texts, replace the "with...Texts." line with this: with the Invariant Sections being LIST THEIR TITLES, with the Front-Cover Texts being LIST, and with the Back-Cover Texts being LIST. If you have Invariant Sections without Cover Texts, or some other combination of the three, merge those two alternatives to suit the situation. If your document contains nontrivial examples of program code, we recommend releasing these examples in parallel under your choice of free software license, such as the GNU General Public License, to permit their use in free software.

MIX
Papier aus verantwortungsvollen Quellen
Paper from responsible sources
FSC® C105338

Printed by Books on Demand GmbH, Norderstedt / Germany